L.A. MATH

L.A. MATH

Romance, Crime, and Mathematics in the City of Angels

James D. Stein

PRINCETON UNIVERSITY PRESS
Princeton and Oxford

Copyright © 2016 by James D. Stein
Requests for permission to reproduce material from this work should be sent to
Permissions, Princeton University Press
Published by Princeton University Press,
41 William Street, Princeton, New Jersey 08540
In the United Kingdom: Princeton University Press,
6 Oxford Street, Woodstock, Oxfordshire OX20 1TW
press.princeton.edu
Jacket art: Detective © Dm-Cherry/Shutterstock
All Rights Reserved
Library of Congress Cataloging-in-Publication Data
Stein, James D., 1941–
L.A. math : romance, crime, and mathematics in the City of Angels / James D.
Stein.
pages cm
Includes index.
ISBN 978-0-691-16828-9 (hardback)
1. Criminal investigation—Fiction. 2. Mathematics—Fiction. 3. Mathematics—
Miscellanea. 4. Los Angeles (Calif.)—Fiction. I. Title. II. Title: Los Angeles math.
III. Title: Romance, crime, and mathematics in the City of Angels.
PS3619.T45L3 2016
813'.6—dc23   2015018419
British Library Cataloging-in-Publication Data is available
This book has been composed in Glypha LT Std & Sabon LT Std
Printed on acid-free paper. ∞
Printed in the United States of America
10 9 8 7 6 5 4 3 2 1

TO
LINDA

—with whom 1 + 1
STILL equals 1

CONTENTS

PREFACE:
L.A. MATH

WHY *L.A. MATH*?

You may be wondering why this is book is called *L.A. Math*. After all, just because math somehow takes place in Los Angeles doesn't mean it's different. $2 + 2 = 4$ here (yes, I live in L.A.), just like everywhere else.

There are actually three reasons. The first has to do with getting people to look at the book. Any time you put "L.A." in the title of anything, you virtually guarantee that people will be interested—I hope, even if the next word is "Math." *L.A. Law* was a successful TV series; *L.A. Confidential* was both a successful book and a successful movie. There's still a mystique and a fascination to L.A. that Wichita, Kansas, and Peoria, Illinois, simply don't have. Even New York, L.A.'s archrival in practically everything, doesn't have it—at least, that's what we in the City of Angels like to think.

The second reason is that this really is a book about Los Angeles and math—although not quite the way you might suspect.

And the third reason? Well, to find that out, you'll have to read on.

HOW THIS BOOK CAME TO BE

I've always loved short stories. A good short story has plot, characters, dialogue—with the added bonus of not having to invest a lot of time reading it. Got fifteen minutes? You can read a short story.

Two of my favorite short story genres are science fiction and mystery. I'm not exactly sure why there seem to be fewer short stories published now than when I was growing up, when many of the greatest authors of science fiction and mystery wrote short stories. I was reminded of this a while back when I found myself in Culver City with a couple of hours to kill. Fortunately, I was near the library and so spent the time happily reading an anthology of the best science fiction stories of 1969. It was time rewardingly spent.

Actually, *very* rewardingly spent, because it was one of the factors that led to my completing this book. I had originally started writing this book more than twenty years ago—using now-defunct word-processing software. I had approached a company that wanted to get into the textbook business with an idea for a different type of text for a Math for Liberal Arts (aka Math for Poets) course. I wanted to write a series of short stories introducing the basic idea for a number of topics that would constitute a Math for Liberal Arts course and then write accompanying textual material that would expand the idea. It would be the most student-friendly math text ever written.

And that's the third reason. *L.A. Math* is an abbreviation for Liberal Arts Mathematics.

The company went for it, they sent me an advance, and I started work. A few months later, the company was taken over by a giant textbook publisher that decided to scrap the project, as it did not fit in with the type of textbook they produced. I had mixed feelings; I wanted the book to see the light of day, but I also wanted to write the book I wanted to write, not the book a panel of educators wanted me to write. I wanted to write a book that would appeal to readers. Whether a textbook appeals to readers is unimportant from the standpoint of textbook publishers; what matters to them is that it will be rejected (for whatever reason) by as few educators as possible.

So I stopped writing the book. I had kept all the files, as well as numerous other files from this period, but I had been unable to convert them into Microsoft Word, which is by now the standard in word processing. I knew I could take the files to a computer expert, pay some money, and have it done—but I'm both cheap and stubborn, and there was no urgency to updating the files, as I had no use for them.

Flash-forward roughly two decades, and here I was, in the Culver City library, reading *The Best from Fantasy and Science Fiction, 19th Series* (Ferman 1971). The first story in the anthology I read was entitled "Gone Fishin'." It was written during the height of the Cold War, so there were some slightly dated aspects, but to say that it was riveting is a giant understatement—it was easily the best short story I had read in a number of years. Memo to any film or TV producer who reads this (and I'm sure there will be lots, LOL)—get the rights, update the story, and make a movie or TV series out of it. I think it's a guaranteed winner. Anyway, the story was written by a Robin Scott Wilson, of whom I had never heard. So, when I got back home, I decided to check him out.

Robin Scott Wilson was a former president of California State University, Chico!

You may not think that this deserves italics, but I did. I teach math at California State University, Long Beach—a much larger branch of the California State University system than the one at Chico, and also, if I may say so, one with more academic substance. In fact, while Wilson was president, CSU Chico was named the number one party school in the country! Wilson took umbrage at this and restored a measure of academic dignity to the campus.

After discovering the not-so-secret life of science fiction author Robin Scott Wilson, I decided to buckle down and see how the stories that I had written for the abandoned math book measured up to "Gone Fishin'." Granted, the genre was totally different, but I've been reading for so long that I can generally tell good writing from bad. You might think that I couldn't be dispassionate about my own stories, but I'm generally my own worst critic. Also, it wouldn't be like I was reading stuff that I had written—I would be reading stuff that somebody else wrote twenty years ago, a somebody else who previously tenanted the body that I now inhabit (regrettably not exactly the same body; the one I now inhabit is somewhat the worse for wear). I'm not the same person I was twenty years ago—who is?

It took me two hours to figure out how to translate the files from the format in which they had been stored by the obsolete word processor. It was embarrassingly easy—but at least I had saved the fee I would have been charged. I started to read the stories. Although I knew the general theme, I had completely forgotten most of the stories—but I was pleasantly surprised, as, I hope, you will be when you read them. I admit that I'm somewhat prejudiced, but I feel that they're all eminently readable, and a few are considerably more than that. IMHO, of course.

MATH CAN ACTUALLY BE ENTERTAINING

Mathematics has managed to infiltrate itself into science fiction. There is an extremely entertaining anthology of mathematically oriented science fiction stories entitled *Fantasia Mathematica*, edited by Clifton Fadiman. Two of my all-time favorite stories appear in it. "The Devil and Simon Flagg," written by Arthur Porges, describes what happens when a mathematician summons Lucifer and bets his soul that the Devil can't come up with a proof of Fermat's Last Theorem in twenty-four hours. The story was written forty years before Andrew Wiles actually solved the problem, but the charm of the story will last as long as there are people to read it. "A Subway Named Moebius," written by A. J. Deutsch at about the same time (the early 1950s), describes the unexpected consequences when the city of Boston constructs a subway system with bizarre topological properties.

These stories, like most science fiction, are one-shot affairs; the characters never appear again in another story. There are science fiction books with recurring characters (think *Star Trek*), but for the most part they are soap operas set in a galaxy far, far away—it is the characters and their interactions that generate appeal, rather than the ingenuity of the stories. Mystery stories are different; it is a combination of the characters, their interactions, *and* the ingenuity of the stories that generates appeal (think *Sherlock Holmes*).

To the best of my incomplete knowledge, nobody has ever tried to do what I started to do twenty years ago: write a collection of short stories, with a continuing set of characters, in which mathematics plays an important role—and a role that varies from story to story. Yes, there was the popular TV series *Numb3rs*; I'm pretty sure that if you liked that series you'll like this book as well.

However, the mathematics in the TV show was sort of a deus ex machina plot device; for the most part, the viewer simply accepted the idea without really "doing the math."

This is different. The stories in this book are unified not only by the presence of a continuing set of characters but also by the fact that the mathematical topics that play roles in the stories constitute a reasonably respectable Math for Liberal Arts course of the type offered by community colleges and universities. You can read the stories and painlessly absorb some interesting and useful mathematics en passant; but if you want, you can plunge a little more deeply into the mathematics by reading the appendix that accompanies each chapter. That portion is written somewhat like a math text—it presents the ideas and illustrative examples relevant to the story, but you'll be happy to know that there isn't a single problem that you'll be asked to do. There is additional material available online at press.princeton.edu/titles/10559.html.

Most first-semester courses in a subject like calculus cover the same material, no matter where you take the course, but Liberal Arts Mathematics is different; it varies a lot from place to place, and even from instructor to instructor at the same school. There are three broad categories of material—different takes on subjects such as algebra and geometry, finite mathematics such as probability and statistics, and relatively recent developments such as game theory and the mathematics of elections. This book includes material from all three categories, so there's almost certainly going to be something in it that is covered in a Liberal Arts Mathematics course if you happen to be taking one.

There are three people to whom this book owes a significant debt. The first is Jordan Ellenberg, who is not only a top-flight mathematician but the author of *How Not To Be Wrong: The Power of Mathematical Thinking*, an absolutely first-rate book about how mathematicians think and why mathematics is so powerful. I had an opportunity to interview Jordan, and in reading his book, I was impressed by how similar his sense of humor was to mine. I asked him to look at a chapter from this book and, if he felt comfortable doing so, to recommend an editor. He suggested the second of the three, Vickie Kearn, who turned out to be the perfect editor for this book. She smoothed out many of the rough edges and anachronisms that were present in the original manuscript during the editing and also helped with suggestions that made the characters more appealing, bringing a sensitivity to the manuscript that it originally lacked. The third is my wife Linda, who has faithfully supported my writing efforts over the years without ever reading a single book I have written—but who has promised to read this one. I'm going to hold her to it. I would also like to thank two people who read the manuscript for their contributions. George Zamba is a detective who brought firsthand knowledge of the detective business to help shore up my admittedly secondhand knowledge of it, and James Coyle came up with several suggestions for improving the stories, including one that put the icing on the cake.

MATHEMATICAL TOPICS BY CHAPTER

CHAPTER 1
Propositions
Logical operators
Truth tables

CHAPTER 2
Percentage

CHAPTER 3
Averages
Rates

CHAPTER 4
Sequences
Arithmetic progressions

CHAPTER 5
Linear equations
Systems of two linear equations

CHAPTER 6
Simple and compound interest
Installment purchases
Amortization

CHAPTER 7
Set theory
Fundamental Counting Principle

CHAPTER 8
Combinatorics
Chinese Restaurant Principle

CHAPTER 9
Probability
Expectation

CHAPTER 10
Conditional probability

CHAPTER 11
Frequency and probability distributions
Mean and standard deviation
Normal distribution
Bernoulli trials

CHAPTER 12
2×2 games
Pure and mixed strategies
Saddlepoints

CHAPTER 13
Voting methods
Arrow's Impossibility Theorem

CHAPTER 14
Algorithms
Traveling Salesman Problem
Task complexity
NP-complete problems

CHAPTER 1

A CHANGE OF SCENE

Santa Monica is up against the ocean. Five miles or so to the east, you'll find Westwood, and in Westwood you'll find UCLA and a lot of movie houses. Between the two is Brentwood, where the rents are more reasonable than Westwood or Santa Monica. That was why I was looking for a place to rent in Brentwood, while I tried to adjust to the fact that, big as New York was, I kept bumping into Lisa when I was there. Really, really awkward. At our wedding, lots of people kidded us about a marriage between a freelance investigator (me) and an artist (Lisa) being an odd-couple type of arrangement. Maybe it was too odd, as we were now separated.

I felt I could use a change of scene, and L.A. is a definite change of scene from New York. At this moment, I was eyeballing a little guesthouse just behind a typical California hacienda off San Vicente north of Wilshire. For those with long memories, that's the general area where O.J. Simpson also had a guesthouse, but O.J. would have turned up his nose at this one—maybe not now, as he was doing a stretch in a Nevada jail, but back when he had money. A rather dilapidated sign declared that it (the guesthouse) was for rent. The sign was dilapidated, but the guesthouse looked okay.

I rang the doorbell and was soon confronted by a casually dressed guy in his late twenties, about six feet two, and a little pudgy. Not everyone in California spends their lives in health clubs. In a pleasant voice with a slight trace of a southern accent, he said, "What can I do for you?" That's one difference between L.A. and New York, where they say, "Yeah, waddya want?" Or just, "Waddya want?" Or just, "Yeah?"

"My name's Freddy Carmichael. I'm thinking about renting your guesthouse," I replied.

"I'm Pete Lennox." We shook hands. He rummaged around and got a key. "It's a nice place, and the location's good." As we traipsed through the main house toward the back, I had my first take on Pete. A sports junkie. Probably a sports bettor. Most of the former over age 18 are the latter, and probably more than a few under age 18 as well, thanks to offshore betting sites.

I didn't have to use any investigative skills to work out Pete's passion for sports, for I'd been in enough man caves. In fact, I'd even had one before I got married. The house was bristling with the latest electronic equipment for viewing, receiving, and recording sports events. A dish antenna on the roof was big enough that it could probably pull in live telecasts from NASA Mars rovers. Tables and chairs were littered with racing forms, sports newsletters, and all the usual paraphernalia that can be found in the home of a typical sports addict. There were mementos of everything from the Kentucky Derby to the Rose Bowl. I saw baseball gloves, tennis rackets, hockey sticks, and basketball jerseys on the way from the living room to the kitchen and out the back door—some autographed, some not.

The environment in which Pete wanted to live was his own affair, but I couldn't help feeling sorry for the goldfish. There were tanks of them all over the place. At least, I assumed they were goldfish because the tanks looked as if they hadn't been cleaned out since Obama was inaugurated. I caught an occasional flash of orange when a fish swam by one of the few places in the glass that wasn't completely clouded over with algae and other things that accrue when an aquarium isn't cleaned regularly. It was a smoggy day in Los Angeles, as it was hot with Santa Ana winds, and the pollutants that had been generated in the L.A. basin had no place to dissipate. Nonetheless, it was a lot better than a hot and muggy day in New York. I had a brief, bizarre thought, wondering if extraterrestrials peering through the gloom for a sight of the Angelenos felt the way I did about the goldfish.

One look at the guesthouse, and it was love at first sight. It had a living room, a den, two bedrooms, two bathrooms, and a kitchen. And an old-fashioned fireplace! I'm a sucker for old-fashioned fireplaces, but I wouldn't have thought that Los Angeles ever got cold enough to need one. There were a few logs, some kindling, and old newspaper to use for tinder, as well as fireplace tongs and a box of really long matches, so I guess it wasn't just for decoration.

I'd have to clean everything up some, but I didn't mind. I wouldn't go so far as to say that I'm a neat freak, but you get a definite feeling of accomplishment when you bring order where before there was only chaos.

A world-class negotiator would have pointed out the guesthouse's defects, but I just wanted to see if it was in my price range. It probably would have been three thousand-plus in Manhattan, depending on where you found it. Of course, there aren't any guesthouses in New York—and this place also had a lawn. And trees. And flowers. There were roses in the garden. Wow! The only time I ever see roses in New York is when I walk by a florist shop—and they cost an arm and a leg if you actually want to buy them.

"How much are you asking?" I inquired.

"Two thousand a month. I'll take care of the utilities."

I may not be a great negotiator, but in New York you *never* take the first offer. "That's a little steep for me. How about seventeen hundred?"

He looked at me, or rather, he studied me. I'd seen eyes like that before. Mostly on Wall Street traders, or gamblers. Neither friendly nor hostile, just assessing what the market will bear. Evidently he thought that the market would bear a little more, for he countered with, "How about splitting the difference?"

"As long as you'll still take care of the utilities."

Either he wasn't out for the last dime, or he just didn't feel like haggling any more. "It's a deal. I've got a contract back at the house." I forked over the traditional first and last month's rent, went to my car (I had rented one for house-hunting purposes), and started to unpack and move in.

Whenever you make a move, it takes life a little while to get back to normal. Your friends and business contacts have to be given your new address, and you have to decide on whether to get a landline in addition to your cell phone. I thought about just keeping my cell phone, but I'm a little bit hard of hearing, and I can hear better with a landline. Besides, there's a certain amount of prestige associated with a Westside 310 area code.

I called Allen Burkitt, owner of Burkitt Investigations, the agency for which I had done some work in New York. I didn't know whether he would have any assignments for me, but I wanted him to know I was available—sort of. I lifted the phone to call Lisa and had punched the "1-212" part of the number when I got cold feet. What if a guy answered? Deciding that if ignorance wasn't bliss, it was at least better than bad news, I handled the problem by sending her an e-mail with my new address and phone number.

I had actually accomplished the impossible by saving some money while living in Manhattan, and so if I wanted to take a few months off, I could certainly afford it. I spent a couple of days acquiring a flat-screen TV, a set of dishes, a new computer, a few prints for decoration, and some L.A. clothes. They dress different out here, and I thought I'd try to blend in. Pete and I had bumped into each other a few times, and he had offered to let me use his spare microwave in the house when he didn't need it. The main house had one of those fancy built-in microwaves, but Pete had a spare microwave and obviously some experience as a single guy. As any single guy knows, the microwave is the greatest of all kitchen appliances, if for no other reason than you can heat leftover take-out food.

Before microwaves, if you had leftover hot and sour soup from a Chinese restaurant, you had to dump it in a pot, heat it, pour it into a bowl (unless you drank it from the pot, which my mother would have frowned upon), and after you consumed it, you had to wash both the pot and the bowl. Now, you just had to heat the microwave-safe container of soup from the Chinese restaurant—which I had just finished doing when the phone rang. It was clearly the house phone; my cell phone has a different ring tone. I left the soup in the microwave to answer my first phone call in L.A.

Maybe it was Lisa! Despite the fact that we were separated, I was still carrying a torch. The adrenaline—possibly mixed with some other biochemical—had started my heart pumping, and my hand was trembling a little as I reached for the receiver.

My adrenaline count went back to normal. It was Allen.

"Freddy?"

"Yeah, Allen. What's up?"

"Welcome to L.A." He paused, but not for long. Allen never forgot that time was money, especially when it was his money. "Listen, Freddy, I could use some help. It shouldn't take more than a day, and there's a thousand in it for you."

I don't know about you, but I'm not in the habit of turning down jobs that offer a thousand dollars for a day or so's worth of work. That's more than two weeks' worth of rent.

"What have you got, Allen?"

"There's a leak in the upper echelons of a corporation. They were about to pick up some valuable West Coast property cheap. You know how the real estate market's been depressed there."

"Yeah, so I've heard. It's depressed everywhere. There's a recession on, in case you haven't heard."

"So they say. Anyway, one of four executives is selling out and is planning to meet a contact from a rival firm out in L.A. and give him specifics of the deal. I've got a man working on the inside who gets pieces of the puzzle. He'll let you know what he's got." He paused for a moment, probably to take a bite of one of those big greasy pastrami sandwiches to which he was addicted.

"And then what?" I prompted him.

"All four executives are heading for L.A. to close the deal. The word is that the one who's selling out will be met by the contact when he or she arrives. I want a description and a photo of the contact, if possible."

"I can guarantee the description. Photos in crowds are iffy, Allen."

"I know it. Just do your best, Freddy. Let me tell you about the possibles."

I grabbed a piece of paper, a pen, and started writing down the names and descriptions of the four executives. Allen could have sent me an e-mail, but he had this thing about e-mail. Someone once told him that the government has a record of everything that goes out, even if you encrypt it, and Allen was old-fashioned enough to think there should still be some privacy in the world. I once told him they probably had a record of his phone calls, too, but he felt it was harder for people to analyze spoken records than written ones.

Anyway, here's the dramatis personae as Allen described them. He told me that Google had fairly recent images of all of them. It helps to have height and weight, though, as from a distance you see outlines rather than faces, and it's hard to get up close to arrivals nowadays with all the security. Of course, you need faces for a positive ID, but occasionally two people with roughly similar faces will be in your field of view, and you can sometimes distinguish them on the basis of such features as height and weight.

Mel Hazlitt. A vice president of some sort. About forty, five-eleven, 180 pounds. Thick black hair, strong jaw, and tortoiseshell glasses.

Don Burns. Head of the legal department. Somewhere in his late fifties, short (five-six) and squat (200 pounds or so). Bald as a doorknob, and not bothering to conceal it.

Vinnie Rossetti. Comptroller (I hated to confess it, but I'm never sure just what a comptroller is). Late thirties, very tall (six-three, 170 or so pounds), receding brown hair. Always looks like he needs a shave.

Elaine Westover. Head of sales. Five-six, 120 pounds, middle thirties (I was impressed—head of sales in her middle thirties). The description Allen gave me was long blond hair, blue eyes (thanks, Allen, you know you never get close enough to see their eyes), smart dresser.

The last item of information was that Arnold, Allen's man on the inside, would call me on Friday with whatever information he had. I was to do the best I could. I always do, I told him.

Friday morning dawned, still smoggy, and I made myself a cup of coffee as I reviewed the pictures I had found on Google. Just as I was finishing the coffee, I got my first call from Arnold. He talked in a whisper, sounding as if he was worried someone might overhear. "Freddy? It's Arnold. Hazlitt is arriving at LAX at about seven this evening, and Burns is coming by train to Union Station. If the contact doesn't meet with Hazlitt, he'll meet with Burns. Got it?" Obviously, he didn't care whether I had it, for he slammed down the phone.

An hour later, the phone rang again. Arnold, whispering even more softly, and even faster. He seemed under a lot of stress. But then, most people in New York are.

"Freddy?" I grunted an affirmative. As soon as he recognized my grunt, he continued, "Westover is arriving at Long Beach around five. If the contact meets with her, he'll meet with Burns as well, possibly to cover up meeting with Westover."

"Then how can I tell which one is selling out?" I asked reasonably. But not sufficiently quickly, for Arnold had hung up.

I made myself another cup of coffee and settled in the armchair by the phone. Shortly afterward, it rang again. Rossetti would be landing at Burbank around six. I could say one thing for Arnold, he didn't spend a whole lot of time on extraneous details.

I wrote down Arnold's latest sound bite and examined my rapidly expanding list of instructions. While I was trying to make heads or tails of them, the phone rang again. Needless to say, I was expecting Arnold. I was wrong.

"Freddy, it's Pete. The built-in microwave just died, and I'm starving. Sorry, but I need my microwave back."

"Sure, I was probably going to get one eventually, anyway. Listen, I've got a little bit of a problem. I'd bring your microwave over to you, but I'm getting some instructions for a job I'm doing, and I need to stay by the phone. Could you come over and get it?"

"I'll be right over." The line cleared. A good thing, too, because almost instantly the phone rang again. Maybe I should get call waiting. This time, though, it really was Arnold.

"Freddy? You remember what I told you about the contact meeting Burns if he doesn't meet with Hazlitt? I got some bad dope, and that's totally wrong."

I was getting writer's cramp. "Yeah, I've got it. But I'd appreciate it if you could simplify things for me."

"No time. I think they're on to me. My battery's low, too." He hung up even faster than before. It didn't sound like I'd be getting any more calls from Arnold.

I contemplated the list of instructions I had, which now covered almost an entire page. What a mess! The four execs had managed to pick locations that were all far away from each other. They were arriving at about the same time, so I couldn't cover all bases. There wasn't any time to try to get anyone else to cover the bases that I couldn't. Finally, I had absolutely no idea who the contact was meeting.

The door opened, and Pete ambled in. I had learned by now that amble was about Pete's usual speed. He picked up the microwave, then came over to where I was sitting and looked at the list of instructions I had compiled. "What's that?"

I briefly outlined the situation. "I can't make any sense of these instructions. Got any ideas?"

He put down the microwave and looked at my sheet of paper for a moment or so. "It doesn't seem too difficult. Just go to Burbank and see who meets Rossetti."

I eyeballed him skeptically. "How did you come up with that, Pete?"

He paused for a moment to organize his thoughts. "Look, Freddy, it's simple if you just look at things right. Either the contact is meeting Hazlitt, or not. However, Arnold's original instructions were that if the contact didn't meet Hazlitt, then he would meet Burns. Therefore, the contact was supposed to meet either Hazlitt or Burns. Or maybe both, but at least one of the two." He paused for breath.

"I'm with you so far. But I still don't see why I'm supposed to pick up Rossetti."

Pete had gotten his second wind. "When Arnold called back and told you that his information was totally wrong, it seems pretty clear that the contact was not going to meet either Hazlitt or Burns. That means that the choice has been narrowed to Rossetti or Westover. Right?"

I nodded. "Right. So now we're down to Rossetti or Westover. How can you tell which?"

"You remember that Arnold said if the contact was to meet with Westover, he would also meet with Burns. But we just agreed that Arnold said the contact wasn't meeting Burns, and so if the contact met with Westover, he would have to meet with Burns as well. That contradicts what Arnold told you. So that leaves Rossetti."

I was convinced. "You've sold me." I was glad that I'd decided to get a car with a GPS in order to reconnoiter the best way to get to Burbank. It may be crowded in Manhattan, but at least it's easy to find your way—even if you can't get there.

Score one for Pete's deductions. Rossetti landed at Burbank, and he was the only tall guy getting off the plane. As the basketball coaches say, you can't teach height. You can't disguise it either. The airport was relatively uncrowded, and I got a couple of shots of Rossetti meeting with a guy in an outfit you certainly wouldn't be caught dead in if you were anyplace outside of California. I returned home, downloaded the pictures, and sent them to Allen. Mission (hopefully) accomplished.

The congratulatory phone call from Allen came the next morning, along with those magic words, "The check is in the mail." When Allen says it, you can trust it, and sure enough, the check arrived a couple of days later. Pete had done me a big favor, and I feel it's bad karma not to return it. What goes around, comes around. I searched online and found a place that would do what I had in mind.

When I arrived at the main house, I found Pete watching a basketball game. Even though it was only two in the afternoon, somewhere on the planet they were playing basketball, and some obscure cable channel in Pete's sports package had located it and decided to televise it. The ads on the signs next to the court might have been in Turkish, for all I could tell. I waited until there was a break in the action.

"Pete, I feel I owe you one."

"What for? Your rent's paid. Did you break anything?"

"Nothing like that. I owe you for helping me to figure out whom the contact was meeting the other day."

He waved his hand. "Oh, that. Simple logic. No sweat." He glanced at the TV, satisfied himself that another ad was upcoming, and continued. "You really want to help me, tell me how the Raiders will do Sunday. They're two-point favorites."

"All I know is what I read in the papers."

"Yeah, but it's nice to know what someone else thinks." He looked at me quizzically. "You follow football, don't you?"

"Yeah, but I'm a Giants fan."

"I might have guessed it. Anyway, do you think the Raiders will cover?"

I shook my head. "No, I think they're going to get killed. But what I really think is that I'd like to help prevent a massacre from taking place here in your house."

"Huh?"

"I've already gotten in touch with Ajax Aquarium Supply. If it's all right with you, they'll come in and clean your fish tanks. I'll pick up the tab. Trust me, you'll like looking at them a whole lot more. And even if you don't thank me, your fish will."

He grinned. "They can't talk. I'll thank you for them."

A short time later the truck from Ajax Aquarium Supply pulled up, came in, and started work. Outside, the Santa Ana had let up, and the smog had lifted, which prompted another one of the bizarre thoughts that occasionally flash across my mind. It occurred to me that maybe the extraterrestrials who might be studying us had decided that their aquarium needed cleaning as well.

CHAPTER 2

THE CASE OF
THE VANISHING
GREENBACKS

The phone rang just as I stepped out of the shower. It was Allen.

"Freddy, are you available for an embezzlement case?"

My biggest success had been in an embezzlement case involving a Wall Street firm specializing in bond trading. Allen had given me a whopping bonus for that one, which was one of the reasons I could afford to take it easy in L.A. I had done well in a couple of other similar cases and had gotten the reputation of being the go-to guy in embezzlement cases. It never hurts to have a reputation for being good at something. Besides, you don't see many guys in my line of work who can read balance sheets.

I've always felt it's important to keep the cash flow positive, and the truth was that I was available for a jaywalking case if it would help the aforementioned cash flow. But it never hurts to play a little hard to get.

"I can probably clear my calendar if it looks interesting."

Allen paused for a moment, either to collect his thoughts or to take a bite of one of those big greasy pastrami sandwiches he loves. "I'm pretty sure you'll find it interesting. It's stumped some people in L.A., and I told them I had a good man out there. I think it's right down your alley."

It's nice to be well thought of—especially by someone in a position to send you business. I knew that Allen's firm, though headquartered in New York, had arrangements with other firms in other cities. I didn't really care about the details as long as the check cleared—which it always had.

"I'm certainly willing to listen. What's the arrangement?"

"Consulting and contingency fee. Fifty-fifty split."

That was our usual arrangement. Burkitt Investigations got a guaranteed fee, plus a bonus for solving the case. Allen and I split it down the middle.

"Okay, Allen, fill me in."

"Ever heard of Linda Vista, Freddy?"

Temporary blank. Movie star? Socialite? Then I had it. Linda Vista was a town somewhere in Orange County with a big art community.

For those of you not up on California politics, Orange County is a bastion of conservatism. You have Orange County to thank, or blame, for Richard Nixon and Ronald Reagan. But Linda Vista, which my fragmentary Spanish translates as "pretty view," was different from your basic Orange County bastion.

The vista in Linda Vista was sufficiently linda that it had attracted a thriving artistic community. There were plenty of artists in Linda Vista, and most of them were liberals.

As a result, Linda Vista was highly polarized. The moderates were few and far between. On the left, you had the artists, with their funky bungalows and workshops. On the right, you had the stockbrokers and real estate moguls, living in gated communities so they wouldn't have to have any contact with the riffraff, except for the tradesmen delivering or repairing stuff. However, there were enough artists and hangers-on to acquire political clout—after all, it's still "one man, one vote" in a democracy, rather than "one dollar, one vote." Pitched battles had raged over practically every issue from A (abortion) to Z (zoning), and many of these battles had made state and even national news.

That's all I knew about Linda Vista, other than not to try to drive down there at rush hour, which turned one hour on the 405 to more than twice that. The obvious question was this: What kind of a contingency case had they got? So I asked it.

Allen filled me in. "The city is out a bunch of bucks, and each side is accusing the other of fraud and embezzlement. Because of the split in the political situation, the city manager gave half the budget to the conservatives and the other half to the liberals, letting each determine how to spend its half. Both sides claim to have been shortchanged."

Allen paused to catch his breath. "I've got a friend who works in the city manager's office. I told him I had a good man out there who'd done a lot of first-class work in embezzlement cases. Want to take a look at it?"

"Sure. How much time should I put in before I throw in the towel?" In other words, how much is the consulting fee?

"As much as you like." In other words, since Allen's meter wasn't running, feel free to burn some midnight oil. "The consulting fee is $3,000, upped to ten if you figure it out and get proof." You don't have to be an expert at division to realize that I was guaranteed a minimum of $1,500 for the time I put in, and $5,000 if I doped it out. You also don't have to be an expert at division to realize that Allen was getting the same amount for making a phone call. I decided to be reincarnated as an employer rather than an employee.

Allen gave me a brief description of the protagonists, and I spent a good portion of the evening with a pot of coffee and my computer, getting some background information on them. I'll say one thing for the Information Age; it's a lot easier to run a background check on people than it used to be. What with search engines and social networks, you save a lot on gas money and shoe leather.

The next morning I waited until after rush hour and then made the trek to Linda Vista. City Hall was located in a section of town where the vista was a long way from linda, unless strip malls filled with 7-11s and fast-food restaurants constitute your idea of attractive scenery. I found a place to park, straightened my coat and tie, and prepared for the interviews.

I was scheduled to have three of them. I had been hoping to arrange for longer interviews, but everyone's in a rush nowadays, and I was getting a quarter-hour with each, tops. They'd all been interviewed previously—Allen had mentioned that this case had stumped others—and people are generally less than enthusiastic about being asked the same questions again. And again. The first interview was with Everett Blaisdell, conservative city councilman, who would explain why the conservatives happened to be short. The next was with Melanie Stevens, liberal city councilwoman, ditto. The last interview would be with Garrett Ryan, city manager.

I have a bad habit. My opinion of members of groups tends to be formed by the members of those groups I have seen before. Consequently, I was expecting the conservative Everett Blaisdell to look like a typical paunchy southern senator with big jowls. So I was a little surprised to discover that Everett Blaisdell was a forty-ish African-American who looked like he had spent years twenty through thirty as an NBA point guard.

He got right down to business. "I want you to know," he barked, "that everything that we have done with our budget allocation has been strictly by the book. Our expenses have been completely documented." He handed me a folder full of ledger sheets and photos of checks, which I glanced at and stashed in my briefcase.

Blaisdell was clearly angry. "The business community is the heart of Linda Vista, and it is ridiculous to suggest that it would act in a manner detrimental to its citizens. We are $198,000 short in our budget."

You don't expect NBA point guards to get out of breath too easily, considering the time they have to go up and down the court, but maybe Blaisdell wasn't in shape. He paused, giving me a chance to get a question in edgewise. "Just what do you think has happened, Mr. Blaisdell?" I inquired mildly.

"I know what has happened. Melanie Stevens and her radical crowd have managed to get hold of that money. They want $200,000 to fund a work of so-called art, which I, and every right-thinking citizen of Linda Vista, find totally offensive. It's mighty suspicious that the missing funds, $198,000, almost precisely cover the projected cost of the statue."

I was curious. "If you don't mind my asking, exactly what is this statue?"

Blaisdell's blood pressure was going up. "They are going to build a scale model of the Statue of Liberty and submerge it in Coca-Cola. You may know that Coca-Cola is acidic, and it will eventually dissolve metal. They say that this so-called dynamic representational art represents the destruction of our civil liberties by overcommercialization. Well, let me tell you, we'll fight it."

He looked at his watch. "Sorry, I've got another appointment. When you find out what those scum have done with the money, let me know." He walked me to his door.

It took a few minutes to locate Melanie Stevens's office, as it was in a different wing of the building, possibly to minimize confrontations between her and Blaisdell. It was a bad day for stereotypes. My mental picture of Melanie Stevens, ultraliberal, was that of a long-haired hippie refugee from the '60s. The real Melanie Stevens was a pert, gray-haired grandmother who looked like she had been interrupted while baking cookies for her grandchildren. She, too, was evidently on a tight schedule, for she said, "Sorry, I can only give you about ten minutes, but I've made copies of all our expenses." More ledger sheets and photos of checks went into my briefcase.

"Let me tell you, Mr. Carmichael, that we could have used that $198,000. We planned to use it for a free clinic. I know exactly what has happened. Blaisdell has doctored the books. I'm sure glad that Ryan had the guts to ask you to look into it."

"Blaisdell seems to think that your people are responsible for the missing funds," I observed.

She snorted. "That's just typical of what they do. Whenever they're in the wrong, they lie and accuse the other side of lying. They rip off the community and channel money into political action committees. Or worse. Blaisdell knows he faces a stiff battle for reelection, and I wouldn't be the least bit surprised to find that money turning up in his campaign fund."

"He seems to think that you are going to use the funds for an art project, rather than a free clinic," I remarked.

"He's just blowing smoke. He knows quite well that the statue will be funded through private subscription." She looked at her watch. "Let me know when you pin the loss on them."

I left Stevens's office for the last interview, with Garrett Ryan, whose anxious expression made it clear that he was not a happy camper. "Have you got any ideas yet?" he asked.

I shook my head. "I've just talked to Blaisdell and Stevens. They've each handed me files containing what they consider to be complete documentation. They've each given me a story asserting their own innocence and blaming the other. I take it that the missing amount is $198,000?"

Now it was Ryan's turn to shake his head. "No, *each* side says that it is missing $198,000. Quite a coincidence. And I'll tell you, Mr. Carmichael, despite the animosity between them, I think that they are both honorable individuals. I find it difficult to believe that either would rip the city off."

I focused on Ryan's coincidence. "It's funny that they are both short exactly the same amount. Perhaps you could tell me a little more about the budgetary process."

"It's really quite simple. Each resident of Linda Vista is taxed a fixed amount. Any complicated tax scheme would just result in a full employment act for accountants. The previous census resulted in a $100 assessment per individual. The population of Linda Vista increased by 20% since the last census. We didn't need any increase in operating expenditures; under my guidance we've done a fiscally conservative and frugal job of running the city. As a result, the council voted to reduce everybody's taxes by 20%. Needless to say, this was a very popular move."

"I'll bet it was. Did everyone pay their taxes, Mr. Ryan?"

"Everybody. We're very proud of that—a 100% collection rate. Despite what you may have heard, the citizens of Linda Vista are very civic-minded. Liberals and conservatives alike."

I've spent enough time with balance sheets to know that accuracy is extremely important. "Was this population increase exactly 20%, or is that merely an approximate figure?"

Ryan consulted a sheet of paper. "Exactly 20%. I have a printout that gives information to four decimal places, so I can be quite sure of that."

Just then a phone rang. Ryan picked it up and engaged in some political double-talk. After a few minutes he replaced the receiver. "Sorry, Mr. Carmichael. I'm behind schedule. Let me know if you make any progress." We shook hands, and I left.

A couple of hours later, I got home, having stopped for a bite but still avoiding rush hour traffic. Pete noticed my presence and asked, "So how'd things go in Linda Vista, Freddy?"

"I had a pretty interesting day. Want to hear about it?"

He nodded. I took about fifteen minutes to describe the problem and the cast of characters. "It looks like I'll have to spend a day or so looking over the books."

Pete shook his head. "It seems pretty clear to me."

I'd seen it before—everybody's a detective. Amateurs always think they know who the guilty party is because it fits in with their preconceptions. I didn't know whether Pete had cast Blaisdell in the role of a political fat cat out to line his campaign war chest or whether Pete was a conservative who saw Melanie Stevens as a radical troublemaker. Anyway, you've got to learn not to jump to conclusions in my line of work.

"You can't do it like that, Pete. You've got to trace down the paper trails. I've done this lots of times."

Pete grabbed a piece of paper, scribbled something on it, and sealed it in an envelope. "Five dollars will get you twenty that the name of the guilty party is inside this envelope."

Pete needed taking down a peg. Maybe two pegs. Besides, I liked getting four-to-one odds on what was obviously an even-money proposition. "You've got a bet," I said. We wrote our names on the envelope, and Pete put it on the table next to the TV.

"Whenever you're ready, we'll unseal the envelope." I headed back to the guesthouse for a session with the books.

Forty-eight hours later, I was bleary-eyed from lack of sleep. I had made no discernible progress. As far as I could tell, both Stevens and Blaisdell were completely on the up-and-up. Either I was losing my touch, or one (or both) of them were wasting their talents, doctoring books for penny-ante amounts. Then I remembered the envelope that Pete had sealed. Maybe he'd actually seen something that I hadn't.

I went over to the main house only to find Pete hunkered down happily watching a baseball game. I waited for a commercial break and then managed to get his attention.

"I'm ready to take a look in the envelope, Pete."

"Have you figured out who the guilty party is?"

"Frankly, no. To be honest, it's got me stumped." I moved to the mantel and unsealed the envelope. The writing was on the other side of the piece of paper. I turned it over. The name Pete had written on it was "Garrett Ryan and the city council"!

I nearly dropped the piece of paper. Whatever I had been expecting, it certainly wasn't this. "What in heaven's name makes you think Ryan and the city council embezzled the money, Pete?"

"I didn't say I thought they did. I just think they're responsible for the missing funds."

I shook my head. "I don't get it. How can they be responsible for the missing funds if they didn't embezzle them?"

"They're probably just guilty of innumeracy. It's pretty common."

"I give up. What's innumeracy?"

"Innumeracy is the arithmetical equivalent of illiteracy. In this instance, it consists of failing to realize how percentages behave." A pitching change was taking place, so Pete turned back to me. "An increase in 20% of the tax base will not compensate for a reduction of 20% in each individual's taxes. Percentages involve multiplication and division, not addition and subtraction. A gain of $20 will compensate for a loss of $20, but that's because you're dealing with adding and subtracting. It's not the same with percentages because the base on which you figure the percentages varies from calculation to calculation."

"You may be right, Pete, but how can we tell?"

Pete grabbed a calculator. "Didn't you say that each faction was out $198,000?"

I checked my figures. "Yeah, that's the amount."

Pete punched a few numbers into the calculator. "Call Ryan and see if there were 99,000 taxpayers in the last census. If there were, I'll show you where the money went."

I got on the phone to Ryan the next morning. He confirmed that the tax base in the previous census was indeed 99,000. I told Pete that it looked like he had been right, but I wanted to see the numbers to prove it.

(Percentage calculation continued on p. 146)

Pete got out a piece of paper. "I think you can see where the money went if you simply do a little multiplication. The taxes collected in the previous census were $100 for each of 99,000 individuals, or $9,900,000. An increase of 20% in the population results in 118,800 individuals. If each pays $80 (that's the 20% reduction from $100), the total taxes collected will be $9,504,000, or $396,000 less than was collected after the previous census. Half of $396,000 is $198,000."

I was convinced. "There are going to be some awfully red faces down in Linda Vista. I'd like to see the press conference when they finally announce it." I went back to the guesthouse, called Allen, and filled him in. He was delighted and said that the check would be in the mail. As I've said before, when Allen says it, he means it. Another advantage of having Allen make the arrangements is that I didn't have to worry about collecting the fee, which is something I've never been very good at.

I wondered exactly how they were going to break the news to the citizens of Linda Vista that they had to pony up another $396,000, but as it was only about $3.34 per taxpayer, I didn't think they'd have too much trouble. Thanks to a combination of Ryan's frugality and population increase, the tax assessment would still be lower than it was after the previous census, and how many government agencies do you know that actually reduce taxes? I quickly calculated that if they assessed everyone $3.42, they could cover not only the shortage but Allen's fee as well. I considered suggesting it to Ryan, but then I thought that Ryan probably wasn't real interested in hearing from someone who had made him look like a bungler.

My conscience was bothering me, and I don't like that. I thought about it and finally came up with a compromise I found acceptable. I went back to the main house.

Pete was watching another baseball game. The Dodgers fouled up an attempted squeeze into an inning-ending double play. Pete groaned. "It could be a long season," he sighed.

"It's early in the year." I handed him a piece of paper. "Maybe this will console you."

"What's this?" He was examining my check for $1,750. "Your rent's paid up."

"It's not for the rent, Pete. It's your share of my fee."

"Fee? What fee?"

"That embezzling case in Orange County. It was worth $3,500 to me to come up with the correct answer. I feel you're entitled to half of it. You crunched the numbers, but I had the contacts and did the legwork."

Pete looked at the check. "It seems like a lot of money for very little work. Tell you what. I'll take $250, and credit the rest towards your rent."

A landlord with a conscience! Maybe I should notify the *Guinness World Records*. "Seems more than fair to me."

Pete tucked the check in the pocket of his shirt. "Tell me, Freddy, is it always this easy, doing investigations?"

I summoned up a wry laugh. "You've got to be kidding. So far, I've asked you two questions that just turned out to be right down your alley. I've sometimes spent months on a case and come up dry. That can make the bottom line look pretty sick. What's it like in your line of work?"

"I don't really have a line of work. I have this house and some money in the bank. I can rent out the guesthouse and make enough to live on. People know I'm pretty good at certain problems, and sometimes they hire me. If it looks like it might be interesting, I'll work on it." He paused. "Of course, if they offer me a ridiculous amount of money, I'll work on it even if it's not interesting. Hey, we're in a recession."

"I'll keep that in mind." I turned to leave the room. Pete's voice stopped me.

"Haven't you forgotten something?"

I turned around. "I give up. What?"

"We had a bet. You owe me five bucks."

I fished a five out of my wallet and handed it over. He nodded with satisfaction as he stuffed it in the same pocket as the check, and then turned his attention back to the game.

CHAPTER 3

A MATTER OF TIME

Under normal circumstances, I gathered that Pete was not averse to houseguests, and I wanted permission to invite a houseguest. However, the circumstances we were under were not normal.

To begin with, we already had a houseguest, Pete's cousin Cindy. Actually, we had two houseguests, for Cindy had brought along Muffy. Muffy (species: dog, genus: miniature poodle) wasn't a person, but if she had been, people would have said that she was spoiled rotten. Muffy's diet seemed to consist of caviar and truffles—at least, she turned up her nose, what there was of it, at generic dog food. Muffy would also yip at all hours, to Pete's consternation, since Pete has a tendency to sleep at all hours. On the credit side of the ledger, at least Muffy was housebroken; otherwise, my chances of getting Pete to take on an additional boarder for a few days would have been approximately zero.

The boarder I was trying to get Pete to take on was Bill Mac-Donald. I guess everybody categorizes their friends, and Bill was on the "old friend, close to but not really a best friend" list. We had grown up in the same neighborhood, gone to school together, and done favors for each other over the years. Bill had become an insurance agent, and I had a case a few years ago in New York that required in-depth research into the insurance business. Bill had supplied it, and I owed him a favor. Bill had done well; in addition to his apartment in Manhattan, he also had an attractive beachfront condo somewhere north of San Diego. Besides, Bill was a perfectly presentable guy who knew enough not to try to sell insurance to his host.

Pete has southern genes, meaning he is by nature quite hospitable. When I asked Pete whether Bill could stay for a few days, he shrugged his shoulders.

"It's your guesthouse. You're paying the rent."

"I know, but I thought I'd run it by you first. No sense antagonizing the landlord."

"What's your friend like?"

"He's an insurance agent. Nice guy, not particularly fascinating, but not a bore. Good manners. Won't mess up the place."

"Tell him he can stay if he doesn't bring a poodle."

Maybe Pete wanted to kill as many birds (visitors) in as short a time as possible. Anyway, I told Bill he was welcome provided he was dogless.

It occurred to me, somewhat belatedly, that I might have given some thought to how Bill would get along with Cindy, and vice versa. I hadn't, so I paused to take stock of the situation.

What I knew about Cindy wasn't a whole lot because she had only been staying with us a couple of days. Cindy was from some place down south, had starred in her senior class play, and had done rather well in drama courses in junior college. Feeling that this was just the ticket to fame and fortune, she had set her sights on a career in the movies and/or TV, and you know where you go when you have set your sights on such a career. Especially when you have a cousin there who can supply you with room and board.

What I knew about Cindy, though, I liked. When you asked her a question, she paused for a few seconds to gather her thoughts and then answered in complete sentences. No pauses for "you know." Maybe it was part of drama class training. Another thing I found endearing was that she was never without a book. I've always liked readers.

To my delight, Bill and Pete hit it off quite well. Admittedly, Pete was already pleased by the fact that Bill hadn't brought a poodle. However, it turned out that Bill was a lot more interested than I was in any activity with a scoreboard, so Pete had someone knowledgeable to discuss sports with. As a result, Pete got the impression that my friends were more than worth talking to. He even went so far as to indicate that invitations to other potential house-guests on my part would be favorably considered.

On the other hand, when Cindy was around, Bill would develop a bad case of incoherent babbling. While Cindy would make complete sentences, Bill couldn't even make sense, much less stick nouns and verbs together. In addition to developing a bad case of brain-lock, his face would become flushed, and one could see beads of sweat break out on his forehead.

Recognize the symptoms? Well, if you don't, I did. They are the unmistakable signs of the onset of lovesickness in the immature adult male. I knew because I had contracted the disease once or twice myself, both as an immature and a mature adult male.

And how did Cindy react to all this? You've got me. I have to say that I didn't think Bill was showing himself to best advantage, but one never knows. Women often say and do things I don't expect— but at least I'm in distinguished company. Freud was reputed to have said on his deathbed, "Women! What do they want?" Freud spent a lifetime studying women. I haven't even spent a third of what I hope will be a long lifetime on the subject, and Freud was undoubtedly a whole lot brighter than I am.

However, I have noticed that some girls prefer guys who are smooth and polished, while others like guys on whom they have an unsettling effect. For Bill's sake, I hoped that Cindy belonged to Category Two, because Bill was definitely unsettled around her.

Meanwhile, Cindy the actress-to-be was doing substantially better than expected. I had heard horror stories of people who went for years in Hollywood without ever landing more than a role as an extra, but within two weeks Cindy actually managed to get a small speaking part in a movie that was due to begin shooting in Chicago in about a week. She had about ten or so lines, which meant that she could get her SAG (Screen Actors Guild) card. Who knows, maybe the producer was stunned to find someone who answered questions in complete sentences and actually read? Anyway, Cindy was ecstatic, and Pete and I were certainly happy for Cindy. As was Bill, as evidenced by the fact that he actually stopped babbling incoherently long enough to tell her so.

An added bonus was that, while three may not always be a crowd, four plus a poodle sometimes makes for an overly crowded household. Pete and I were both looking forward to regaining sole possession of the premises because Bill was scheduled to leave at about the same time that Cindy left for Chicago. To do some rehearsing on the way, she was planning to take a train and familiarize herself with the script.

Bill had taken to confiding in me his feelings for Cindy, not realizing how unnecessary it was. I didn't have to be a genius to realize that he was hoping that I would intercede somehow to get him a date with Cindy. He didn't ask me directly, but it was clear what was going on. As far as I was concerned, nothing doing. I had gotten him room and board, and enough was enough. If he had been on the best friends list, I might have been willing to take a shot (although this can often be an easy way to lose a friend, even a best one), but as things stood, not a chance. Besides, even if I had been willing, I didn't know Cindy well enough to chance a direct encounter, and doing it indirectly through Pete was clearly inadvisable. You don't ask your landlord to fix his cousin up with your friend. At least, I don't.

Of course, I wished Bill the best of luck, but I felt that it was up to him to work things out in the romance department, and on the eve of Cindy's departure, it was clear that he had yet to do so. On awakening the next morning, I found that I had run out of coffee. I can't function without my morning coffee, and fortunately Pete's sense of southern hospitality extends to items such as this. Bill and I traipsed over to the kitchen of the main house to remedy this deficiency. There I was, at about 8 a.m., making coffee for Bill and myself, when I suddenly heard one of the most anguished sounds ever to split the silence of a tranquil morning.

It woke Pete, and we both charged into the living room, the source of the aforementioned anguished sound. I saw Bill with a rolled-up newspaper in his hands, about to deliver the coup de grace to Muffy. Dogs like to chew things. Muffy had somehow managed to masticate all but a few pages of Bill's customer account book, which he had foolishly taken into the main house in range of Muffy. I say "foolishly" with the benefit of hindsight, which has an unblemished track record of being notably more accurate than foresight.

I stopped Bill in midbackswing. I should note that it wasn't clear whether the aforementioned anguished sounds came from Bill or Muffy.

"I don't think pounding Muffy to a pulp is too likely to score points with Cindy," I said to Bill.

He looked crestfallen, although I could not say for sure whether the crest had fallen because of the destruction of his customer account book or because he could not properly work out his aggressions against Muffy. I suspected that Bill was jealous of Muffy because Muffy clearly had a place in Cindy's affections, whereas Bill was still on the outside looking in.

After a moment, however, he came back to business. "Freddy, I need a large favor. Can you drive me to my condo? I've got a copy of my account book at home, and I need it tomorrow."

I was less than enthusiastic, but there wasn't anything on my calendar for that day. Nonetheless, I didn't relish a day's worth of driving. "Can't you get someone to fax it to you, or something? Don't you have it on your computer?"

"Yes, but it's encrypted, and I don't leave the decryption code on the computer. It's in the house, too, and I've got the only key."

"You could rent a car," I proposed.

"True, but I'd like to get back in time to say good-bye to Cindy. Renting it, driving to the car rental place, and picking up the car would take a couple of hours."

"What about taking a train?" I was still looking for a way out. "The gridlock on I-5 gets worse and worse."

"Trains don't run often enough for me to get back in time."

I guess I could have let Bill take my car, but I realized that sacrificing a day of my life now would be a good idea. As I said, Bill had done very nicely in the insurance business. Should I decide to visit New York, it would be nice to have a place to stay—especially considering that any place in New York that passed a sanitation code inspection cost an arm and a leg. A few hours invested now could obviously pay dividends down the road. Besides, he was a friend.

"Okay, I'll do it."

"Thanks, Freddy, I owe you one. Let's see, my condo is about 120 miles away. It's morning now, so we should be able to average about 40 miles per hour on the way down, and it won't take me more than a couple of minutes to dig up my account book. Unfortunately, I read that today they're going to shut down a couple of lanes of northbound I-5, and we'll be lucky to average 20 miles per hour on the way back. What time does Cindy have to leave to catch her train?"

"Somewhere around 5:30." I calculated rapidly. "Forty plus twenty divided by two is thirty. We'll average thirty miles per hour. That's eight hours for a 240-mile round trip. If you hurry, we should be ready to leave a few minutes before 9:00. That should give us enough leeway."

"We have to eat. I have to, anyway."

"Not to worry. There's undoubtedly something in the fridge we can take with us, or we can stop for a Big Mac." I turned to Pete, who was not thrilled about being awakened by anguished sounds, a subparagraph under the general heading that Pete is not thrilled about being awakened.

"Pete, do you suppose you could fill Cindy in while we're on the road?" I should probably mention that Cindy, who slept with earplugs (probably to drown out Muffy's early-morning yipping), was blissfully unaware of developments.

"I'll leave her a note." Pete yawned. "I only got to bed at 3:00 a.m., so I'm going back to sleep." Yawning some more, he departed. I told Bill to leave a note as well, as I had no idea when Pete would awaken.

Bill and I got dressed hurriedly and were on our way a little before 9:00. Bill was obviously familiar with freeway conditions, and as he predicted, we were able to average about forty miles per hour on the way down to San Diego, arriving a little before noon. It took only a few minutes for Bill to grab his duplicate account book, and we were once again on the road.

The return trip to Los Angeles was a nightmare, as it always is when two lanes of a major interstate are shut down. Nonetheless, we were able to manage almost a constant speed of twenty. You can see why I tried to duck out by steering Bill toward Amtrak.

The afternoon wore on. And on and on. What made it even more wearing was the fact that Bill spent the afternoon discussing romance, women in general, and Cindy in particular. As a philosopher, Bill left much to be desired. After a couple of hours of this, I managed to turn on the radio to a Dodgers game. I'm not really a baseball fan, but Bill is, and I'd rather listen to live baseball than a rehash of the past couple of hours of Bill's angst.

Finally, as we neared the outskirts of Long Beach, I took a glance at my watch, and was horrified to find that it was almost 5:00! What had gone wrong? We had averaged a little better than forty miles per hour on the way up, and almost exactly twenty miles per hour on the way back.

We arrived back at the house a few minutes after 6:00. Pete greeted us and said that Cindy had left a little over half an hour ago. Bill cursed.

"I thought you said we'd be back in plenty of time, Freddy."

"I thought we would. I don't know what went wrong."

Pete had been lounging on the couch but had obviously been paying attention to the conversation, for he turned to face us and said, "I think I can tell you what happened."

Bill was still glowering at me. Pete might have been annoyed by having been awakened early, but at least he wasn't glowering at me. Deciding that it was better to converse with a nonglowerer, I told Pete, "I don't get it. We averaged forty miles per hour on the way down and twenty on the way back. We only needed a couple of minutes to get Bill's account book, and another five to grab a burger and fries. We should have had time to spare."

(Calculating averages continued on p. 150)
"You assumed that the average of twenty miles per hour and forty miles per hour is thirty miles per hour. That is only the case if you drive for equal *times* at those speeds. The two of you, however, drove for equal *distances* at those speeds. You actually took three hours to drive down and six to drive back, an hour more than you thought you would."

Bill glared. "This is all your fault, Freddy."

"My fault? I get you room and board. The girl of your dreams is under the same roof. You, on the other hand, leave your account book where it can be chewed by dogs. I take an entire day to chauffeur you, and this is the thanks I get?" I was more than a little angry.

Bill backed off. "Sorry, Freddy. I should have asked her for a date sooner, but I just didn't have the guts."

"Why not ask her for a date now?" Pete suggested.

I won't say that Bill's face was suffused with the glow of hope, but the glower diminished in intensity. "What have you got in mind?"

"Why don't you take care of your insurance business and then fly to Chicago? You can meet her when her train arrives."

Bill thought it over. "Yes, I could do that. But I don't even know how she feels. I could end up with an awful lot of egg on my face. Besides, she may have made other plans, and she certainly won't be expecting to see me."

Pete gave me a "do-I-have-to-fill-in-all-the-details-for-him?" look. Well, he gave me a look, and that's how I decoded it after he said, "You must have some idea of what she likes. Roses, chocolates, Chanel No. 5—something along that line. Have it delivered to her stateroom. Trains make stops along the way, so you can certainly arrange this. Sign the card "From a secret admirer." Then, *after* the receipt has been acknowledged, send her a text or an e-mail revealing that you are the secret admirer. Say that you had business in Chicago and you'd love to get together with her. It's done all the time."

Bill brightened. "It's worth a shot. Besides, Chicago is the home office. I could always charge the trip off to business." I wondered if he would charge the roses (or chocolates or Chanel No. 5) off to business as well.

This was the first time I had ever seen Pete venture into the relationship realm, even if it was someone else's relationship. It was apparently a successful venture, as he showed me a selfie that Cindy had attached to an e-mail a few days later, showing Bill and Cindy happily walking on the Chicago lakefront. As I was looking at it, a question suddenly popped into my head.

"Pete, that little calculation you did with averages wasn't very complicated. How come you didn't realize that we'd be late for Cindy's departure? Did you figure it out after we left? You could have stopped us from going."

I hadn't known Pete all that long, but I had learned that nothing ticks him off more than being accused of having failed to figure something out, even when his brain is only working at half-speed due to lack of adequate sleep.

"That little problem?" he snorted. "I knew what the story was the moment you assumed that the average of twenty miles per hour and forty miles per hour was thirty miles per hour. High school stuff."

"Then, if you knew we would be late, why didn't you stop us?"

"I had no reason to stop you. Besides, I had problems of my own," Pete declared. "Bill may be your friend, but Cindy is my cousin. She wanted to know how Bill felt about her, and I had no idea. You weren't exactly Dear Abby."

I looked at him. "You couldn't tell how Bill felt? It was obvious."

"Maybe to you. I didn't have a clue."

"Well, it seemed clear enough to me. I must admit, though, Bill wasn't about to take any action until he arrived too late to say good-bye."

"It reminds me of how I used to handle things when I was a kid and didn't know how I felt," Pete recalled. "When I wasn't sure what I wanted to do, I would flip a coin to help me decide. As soon as the coin landed, I knew which way I was rooting for it to land."

I nodded. "I see. As soon as he saw that Cindy had gone, possibly for good, Bill realized that it is better to ask for a date and get turned down than never to have asked at all."

"That's the way it looks to me."

"Well, Pete, you've made two people happy. Bill and Cindy. At least, so far." Suddenly a thought struck me. "This situation probably was a first for you, Pete."

"Making two people happy?"

"I wasn't talking about that. As I see it, if you had announced that we wouldn't get back in time, Cindy and Bill might never have worked things out. Bill might have continued to babble incoherently, and the way he was going, he would probably never have been able to ask Cindy for a date."

Pete thought about it. "Maybe so. So what?"

"Well, you would have achieved the same results if you *hadn't* solved the problem of when we would get back. Once you solved it, there was a risk you might screw everything up by announcing the solution. So it's probably the first time that you would have made out better by not solving the problem in the first place."

He thought this one over. At least, he gave the appearance of thinking this one over. What emerged was, "Not true. Did things work out? Yes. Would they have worked out otherwise? Who knows?" He then did what I always do when I feel that my debating position is shaky. He turned and left.

CHAPTER 4

THE WORST
FORTY DAYS
SINCE THE FLOOD

I suppose it was only a matter of time. After all, Lisa was in New York, and as someone once said, there are plenty of other fish in the sea. My first attempt to cast a line for the aforementioned other fish came about by accident. I had run out of edibles, and so I decided to go to the supermarket and restock. You can always find an all-night supermarket in L.A. in your neighborhood. You can always find one in New York as well, but the selection is more limited because space is at much more of a premium in New York. Also, despite the falling crime rate in big cities, there's more chance of running into danger in Manhattan than in Brentwood.

One of my failings is that I don't always look where I am going. As a result, when I walked into the only open door at the supermarket, my mind was elsewhere, and I bumped into a young lady who was just emerging. Since I outweighed her by thirty to forty pounds, and since she was carrying groceries and I wasn't, she suffered substantially more damage in the collision.

"Why don't you watch where you're going?" she snapped.

In New York, a remark like this is a prelude to a pas de deux in which the other party is obligated to make some remark like, "Watch it yourself, dummy!" In fact, a remark like this was on my tongue. Then I took a second look at the young lady, which made me realize that this might be a good occasion for the soft answer that turneth away wrath. Besides, this wasn't New York. So I swallowed my pride, apologized, and asked if I could pay for whatever damage I had caused.

She looked stunned. "Nobody ever offered to do that when I lived in New York!" she exclaimed, looking through her shopping bag to see what damage I had caused.

Well, even I couldn't mess up an opening like that—especially since I followed through and replaced the carton of eggs in her shopping bag, one of which was cracked. Further conversation disclosed that her name was Erika Nussbaum, that she had come from New York to Hollywood to become a model, and that she had just gotten off a modeling assignment for a face cream that others may have needed but she certainly didn't. More to the point, yes, she would appreciate grabbing a bite to eat at the all-night coffee shop next door. One thing led to another, and soon I had my first date in seven years (marriage doesn't count as a date).

Let me pause to insert a remark or so about myself. Like most people, I have a pretty reasonable opinion of myself, but I recognize that I have my fair share of failings. As a result, I tend to look tolerantly on the failings of others. Let he who is without sin cast the first stone.

One of my failings is a liking for tobacco. Actually, I do not feel that this is a failing. In the 1930s and 1940s, smoking was a sign of sophistication. Not only am I living in intolerant times for lovers of tobacco, I am living in an intolerant place. Los Angeles seems bent on ridding itself of smokers with the same messianic zeal used when Salem tried to rid itself of witches. Everybody would be a lot better off if Los Angeles devoted itself instead to getting rid of graffiti.

Speaking of being a lot better off, I might have been better off if I had met Erika Nussbaum through an Internet dating service because she would have inserted the phrase "n/s" in it, which means that she wanted a nonsmoker. As you will see, this would have saved a lot of grief.

A paragraph or so back, I was talking about failings. I am sure that Erika, like everyone else on the planet, had her fair share of failings. Failings, yes. Flaws, no. Erika may have hailed from New York, but she looked like a California surfer, with long blonde hair, clear blue eyes, and a perfectly shaped nose. I don't know why, but I have this thing about perfectly shaped noses.

It didn't take long for me to discover that Erika shared the attitude that Los Angeles had about smoking. On our second date (not counting the late night encounter at the coffee shop), we had finished dinner in this nice little Italian restaurant I knew (one of the few remaining that allow you to smoke) and had ordered coffee. The coffee arrived, and I automatically lit up my after-dinner cigarette. Erika's blue eyes narrowed, and the perfectly shaped nose wrinkled in distaste.

"Freddy," she said, in a tone which also indicated distaste (as if additional evidence was needed), "you simply must make an effort to give up smoking. It's a disgusting habit. It's not only unhealthy for you; it's unhealthy for those around you."

Trained detective that I am, I had no trouble picking up on the distaste indicated by her tone, reinforced by the narrowing blue eyes and the wrinkling nose. "Erika, it's not as easy as that. You speak as a nonsmoker. I ask that you remember the words of Mark Twain. 'It is easy,' he said, 'to give up smoking. I have done it thousands of times.'" I chuckled slightly, hoping that this light remark would steer the conversation away from troubled waters. No luck.

The blue eyes narrowed further, and the nose wrinkled even more. "Freddy, I cannot see myself being continually exposed to harmful carcinogens. If you cannot give up smoking immediately, then do it gradually. Cut down on a day-by-day basis."

I didn't like this one bit. On the other hand, I wasn't going to give up a burgeoning relationship without a fight, even if it was against an ingrained habit.

"All right," I relented. "I'll do it. I smoke two packs a day. That's forty cigarettes. I'll cut down one cigarette a day for the next forty days. How does that sound?"

Erika's eyes widened slightly. "That's wonderful, Freddy." They widened even more, and the nose resumed its original shape. Her voice, too, softened attractively. "Since we're almost through with dinner, I have a suggestion to make." Her hand reached out for mine.

I liked the way things were going. For about eight-tenths of a second.

Her hand briefly caressed mine and then reached its intended target, my cigarette. Deftly she removed it from my fingers and stubbed it out. "Why don't you start cutting down right now?"

The evening did not end quite the way I had hoped. But I had a job to do, and my cell phone calculator seemed the ideal tool with which to do it. I wanted to calculate how many cigarettes I would need to complete the program upon which I had promised Erika I would embark.

However, there was a problem. Have you ever tried to add up the numbers 1 to 40 on your cell phone? Not only did it threaten to take some time, I made mistakes keying in the numbers. I thought about waiting until tomorrow, but I noticed a light on in the main house. I thought maybe Pete might have a solution.

"Pete, I could use some help. I've got a math problem, and I thought it might be in your ballpark."

Pete actually *liked* math problems. "I'll take a shot at it," he replied eagerly.

"I don't really need to do anything complicated. I just have to add up the numbers from 1 to 40. Erika talked me into quitting smoking by reducing from my habitual two packs a day, by one cigarette per day. I'm out of cigarettes, so I thought I'd calculate exactly how many I need, and then go out and buy them."

Pete's answer came just nanoseconds after I finished. "Just buy four cartons and one extra pack."

I looked at him quizzically. "You're bluffing. I call. No human being, and very few computers, could add up the numbers from 1 to 40 in that short a time. I don't want to run short of cigarettes, nor do I want to have any left over."

"Who's bluffing?" The only times I have ever seen Pete get upset are when he thinks you think he's wrong, and he feels he's right. "Just for the record, I did not add the numbers from 1 to 40. I computed their sum."

"What in heaven's name is the difference?"

/ (Number of cigarettes needed continued on p. 158)

"Look, Freddy. One and 40 total 41. So do 2 and 39, 3 and 38, and so on. The last such pair is 20 and 21. There are consequently twenty pairs, each totaling 41. By coincidence, each pack of cigarettes contains twenty cigarettes. So you need 41 packs of cigarettes. Since there are ten packs in a carton, you will need four cartons and one extra pack."

I digested this. "Right you are. Could I ask you to dispose of my cigars for me? They're Havanas, and I guess it's probably best to remove temptation in all its forms."

"Sure enough. I'm going to put some pizza in the microwave. Want any?"

"No, thanks. I just had Italian for dinner."

He came back in a moment, happily munching away. "You must be very interested in Erika to contemplate such a drastic change in lifestyle. If you don't mind my asking, how are things going?"

"Good question. I never know whether it's a good sign or a bad one when they want to remodel you."

Pete finished the last of his pizza. "I know what you mean. Anyway, I hope things work out." Before I lost my resolve, I went to the guesthouse and got the cigars. As I handed them to Pete, I felt like I was losing a friend.

In following this tale to its conclusion, I have decided to divide it into quarters, like a football game. During the first ten-day quarter, I reduced the number of cigarettes I smoked from 40 to 31. I therefore smoked a total of 355 cigarettes during that interval. Forty plus 31 is 71, as is 39 and 32, etc. There are five such pairs, making a total of 355 cigarettes. I may not be quite as good at calculations as Pete, but this particular trick was actually child's play, as he had pointed out.

I would have thought that I would have started to experience some measure of withdrawal symptoms, but such was not the case. Possibly this was due to the fact that what I was losing in tobacco I was gaining in time spent with Erika. Of course, this had been my reason for quitting smoking, but it was nice to see immediate dividends. Pete met Erika but did not seem as enchanted with her as I was. I suspected that this was because Erika devoted the majority of that meeting to a discussion of the perils of the pizza and beer Pete was wolfing down. Or maybe he just didn't appreciate perfectly shaped noses as much as I do.

Second ten-day period: reduction of cigarettes from 30 down to 21. In this interval, I smoked a total of 255 cigarettes. Since I wasn't smoking so much, I also had a little additional leisure to notice things. One of the things that I noticed was that 255 was 100 less than 355. It occurred to me that, on the first day of the second period, I was smoking 30 cigarettes, which was 10 less than the 40 cigarettes I had smoked on the first day of the first period. On the second day of the second period, I was smoking 29 cigarettes, which was 10 less than the 39 cigarettes I had smoked on the second day of the first period. And so on. Since I was smoking 10 cigarettes less per day for each of the ten days of this period, it was perfectly natural that I should be smoking 10 times 10, or 100 fewer cigarettes during the second period than I did during the first.

I was beginning to believe that stuff about tobacco dulling the brain. Mine certainly seemed to be becoming more agile.

I confess that I was noticing some other things as well. Erika and I were dining one evening when she said, "Freddy, why do you keep looking at your watch?"

"I'm sorry, angel," I replied. "It's not as easy to cut down on smoking as I might have thought, so I have put myself on a schedule. I can smoke my next cigarette in, let's see, fourteen minutes."

"I really am pleased that you are making such a determined effort to see this project through, Freddy. And, I must say, schedules are a very good way to do it." She paused. "Which reminds me. Perhaps you should adopt a healthier diet as well. Your diet is terrible. Hamburger, for instance, is just loaded with cholesterol. I'll get the nutritionist at the health club to work up a diet for you."

I looked at my watch. Twelve minutes to go. I looked at Erika. Yes, the nose was still everything a nose should be, but I hadn't really noticed the way her jaw sometimes acquired a hard set when she thought you might be opposed to one of her ideas. Ah, well. I guess, when you are escorting a girl whose face, like Helen of Troy, would launch a thousand ships under optimum conditions, you must be prepared for moments when they would launch only, say, 930. Absent-mindedly, I reached for a cigarette.

"Freddy!" Practically a screech. "I'm sure you have *at least* ten minutes to go before your next cigarette."

I recovered nicely. "Time just seems to fly when I'm with you, Erika." I replaced the cigarette and looked at my watch. The second hand just seemed to crawl.

Third ten-day period: reduction of cigarettes from 20 per day to 11 a day. By now I was noticing some definite side effects. Food, instead of tasting better (as all the quit-smoking experts will tell you), was tasting worse. The first intimation of this came one day during lunch. Pete had talked me into visiting a Mexican restaurant he had just discovered over on Pico. I ordered nachos, chiles rellenos, and beef enchiladas. The nachos seemed okay, but I could have sworn there was something wrong with the chiles rellenos. They tasted sour. Since bad food can put me out of commission for several days, I decided to get a second opinion. "Pete, would you mind tasting these rellenos?"

Pete took a couple of bites, chewing reflectively like an inspector for the Guide Michelin considering the merits of tournedos Rossini. "Yeah, I see what you mean. Not hot enough."

"It doesn't taste sour to you?"

"Just a little flat." He added a couple of jalapeños, squirted some Tabasco , and then sampled again. "Try it now."

I did. It tasted a lot hotter, but still sour. I said as much, recalling that one of Pete's attributes was a stomach with an interior like a blast furnace.

"You know, Freddy, nicotine affects the taste buds. I'm sure that's what happening." Maybe he was right.

The other thing that I noticed was that both friends and acquaintances were becoming rather snappish. Even Pete seemed to be developing a short fuse of late.

I was also becoming less and less enchanted with Erika. It's funny, but I hadn't noticed how insistent she was on *always* having her way. Not only that, but when I even dared to suggest that her plans for creating a new Freddy Carmichael might be a bit extreme, I started hearing comparisons between myself and a paragon of virtue named Stefan. Or something like that. Quite frankly, I was getting heartily sick of it.

Fourth ten-day period: reduction of cigarettes from 10 to 1. Every day, the first order of business was to calculate the interval between cigarettes. It seemed to be lengthening alarmingly. I called Pete's attention to it one morning.

"What did you expect?" he growled. "If you reduce the number of cigarettes from 10 per day to 9 per day, the interval between cigarettes will increase reciprocally. Assuming that each day is the same length, the interval between cigarettes will increase by 1/9 today, or about 11.1%. Tomorrow it will increase by 1/8, or about 12.5%."

"I want sympathy, not a lecture in mathematics."

"She's your girlfriend, not mine. If you ask me, you'd be a lot better off without her."

See what I mean? Everybody was turning into a grouch.

The climax came on the day when I was down to just three cigarettes. I had decided to have them with coffee after each meal. Erika and I had finished dinner, and I lit up. So did her eyes. Pigs' eyes are also blue.

"Freddy, haven't you quit yet? Stefan doesn't smoke. He also is not overweight. How are you keeping to your diet?" She surveyed me critically. "Your muscle tone could be a lot firmer. You really must exercise more."

By now, I was spoiling for a fight. "Erika, in the last few weeks I have given up tobacco and substituted tofu for hamburger. What I may have gained in health I seem to have lost in joie de vivre. It's not how long you live, but how much you enjoy it."

I took one more look at that perfectly shaped nose, probably for the last time, inhaled a good deep lungful of smoke and savored it. It was more fun than I had had in weeks.

Erika took a look at me, almost certainly for the last time, then shook her head and got up from the table and left. I wasn't even remotely tempted to follow.

I told Pete about it that evening. "And so," I concluded, "Erika and I are no longer an item. I'm guessing that Stefan, a nonsmoker, has replaced me in Erika's affections. Win some, lose some."

"I think his name is Stefan Ericson. He's a ski instructor."

I looked at Pete. "And just how do you know this, if I may ask?"

"One evening a few weeks back, you may recall that the three of us went out to dinner. While you were away from the table, Erika said she was planning on going skiing at Mammoth and wanted to know if I knew any instructors. I recommended Stefan."

"Oh, you did, did you? Did you have any idea that this ski instructor was going to rip off my girlfriend?"

Pete countered with a question. "Just a moment here. Weren't you just telling me a few moments ago that *you* told *her* off?"

"Yes, but that's because the best defense for an about-to-be-bruised ego is a good offense." I leaned back in my chair. "I guess things worked out for the best. I'm sure it would be better for my health if I went back to my old habits. At least, for my mental health. I'm young, reasonably fit, and I'll worry about my lungs later." I took a deep breath. "You know, I sure wish I hadn't asked you to dispose of those Havanas. One would really taste great right now."

Pete went into his bedroom and emerged some thirty seconds later. With my Havanas! "I didn't get rid of them, Freddy. I had a hunch you'd want them back."

As I lit up, I could feel peace and tranquillity descending once more, a sensation I never felt with Erika. To be completely fair, Erika was capable of arousing feelings that you just don't get with cigars. Reflecting on this, I recalled that Rudyard Kipling once said that a woman was only a woman, but a good cigar was a smoke. Maybe it was also Rudyard Kipling who had observed that there were other fish in the sea.

CHAPTER 5

THE ACCIDENTAL GUEST

Having spent most of my life in New York, I've had little experience with car rental agencies. To be perfectly frank, I've had little experience with cars. In New York, if you want to get some place, you take a taxi, a bus, or the subway. Or you walk.

However, it's almost impossible to survive in Los Angeles without a car. Sure, you can do it if you have to, but because L.A. is much more spread out than New York, a car becomes almost a necessity. Especially if you want to have any kind of a social life, no matter what your definition of "social life." So I had purchased a late model used car, which was serving me quite well until Thursday morning. I had been sitting at a stoplight, happily minding my own business, when the car behind me slammed into me. I was all right, but the car had suffered a few more-than-superficial dings, the rear axle was out of alignment, and there was damage to some of the key components that enable the car to move. At least, that's what the mechanic in the shop said—and looking at the car, I tended to believe him. Plus, I had to get it towed to the shop because when I turned the key in the ignition, absolutely nothing happened. It would take a couple of days to get it fixed.

At least it had happened near home, so although the placing of the accident was pretty good, the timing couldn't have been worse. I had been invited to spend the weekend at Carl and Peggy

O'Hara's ranch, out by Santa Barbara, more than a hundred miles away. They were having a weekend gathering of some friends and had even printed up invitations, including a map—they lived out in the country, where the roads aren't always clearly marked, the way they do in L.A. with overhead signs. In the country, often the roads are marked with these little dirt-brown small pedestals on the roadside on which the name was once painted but has now faded. Thank goodness for GPS! Even though Pete didn't have any plans for the weekend, it would have been a major imposition to ask to borrow his car for several days. So I did the obvious thing and located a car rental agency within walking distance. They had a choice of several different cars and plans, and so I opted for a subcompact for $60 for the weekend and 15 cents a mile. I wanted to save a few bucks, and the only things that would be going into the car were me and my overnight bag.

I'd heard that the California coastline between Los Angeles and Santa Barbara was well worth seeing, especially the Channel Islands, so I decided to make a day of it and departed bright and early Friday morning. The Channel Islands were certainly different. I've been to Jones Beach, but I didn't see spectacular stone arches in the Atlantic, or beaches overflowing with seals, or kelp forests. I'm glad I took the time for the detour.

Looking back later over the weekend's events, it was extremely lucky that I had indeed decided to leave early. I had Pete to thank for that, as he told me that Friday afternoon traffic heading north on the 101 turned a normal ninety-minute trip into one twice that long.

I must admit, though, when I pulled up late that afternoon at the O'Hara ranch, I did not think that I had been extremely lucky. Quite the opposite. There had been a terrific downpour Thursday night, and the access road to the O'Hara ranch was a sea of mud. The car got stuck in it a couple of times. I had to get out and dig out some of the mud, using only my hands and the few tools I could find in the emergency kit. Ever try scraping away mud with a tire iron? Consequently, when I rang at the front door, I was badly in need of a shower.

Peggy greeted me at the door. "You must have left pretty early this morning, Freddy."

"About eight o'clock."

"Yeah, we tried to reach you at 8:30, but you had already gone. We're advising all our guests to take the train to Santa Barbara. We're used to conditions like this, and Carl has a heavy-duty off-road vehicle that can fight its way through anything. Sorry I didn't get to you in time."

"Something happened to my cell phone, and I took it into the store. I guess I'm stranded in the previous century for this week. Oh well, it was an interesting drive up, and no real harm done, as long as you've got a place where I can shower and change."

She looked me over. "Maybe I should have Carl pick up some extra soap on his next trip." We both laughed, and I followed her up to an attractive guestroom on the second floor. Half an hour later, I was downstairs, prepared to greet my fellow guests.

I didn't have to be a detective to figure out when they were arriving. The O'Hara's off-road vehicle may have been reliable, but even I could tell that it was badly in need of a tune-up. I could hear the occasional misfire and engine knock when it was more than a mile away. Maybe the car wasn't in bad shape, though—when you're used to the city noises, the silence of the country seems to make your hearing improve.

The O'Haras had corralled an eclectic collection of weekend guests, so I'll list them in descending order of interest (to me). I would certainly have put Ann Robinson, a lifelong friend of Peggy's, at the top of the list. We had some things in common—I was separated from Lisa, and Ann was recently divorced. I gathered that Ann's presence was somewhat unexpected. Peggy had issued the invitation, and Ann had originally declined, saying that she was going to Cannes for a couple of weeks. It turned out that her departure for Cannes had been delayed for a few days, and here she was, with enough clothing and jewelry to sink a ship. I must admit I was a little nervous with Ann at first because most of my friends tend to spend their vacation weeks in upstate New York as opposed to the south of France, but after a while, we got along just fine.

Myron Wallace wasn't exactly my cup of tea, but I had to admit he was an interesting guy. He was a food and restaurant critic for some highbrow magazine. I like food as much as the next person, but to Myron, it was a way of life. I guess one of the reasons I found Myron interesting is that I always enjoy expertise. In my line of work, the more you know about lots of things, the better. After a weekend with Myron, not only had he taught me a lot about haute cuisine, he had also given me the names of a couple of pleasant and inexpensive French restaurants in Los Angeles that he had visited incognito. When you are six feet four, bald, and have a goatee like Myron Wallace, I wonder how you can visit restaurants incognito, but maybe there is a legion of Myron Wallace look-alikes making the rounds of restaurants. I looked forward to checking out his recommendations.

Sue Fredericks was an editor. Ordinarily, I would think that an editor would be more interesting than a food critic, as the editor would be meeting writers, whereas the food critic would be meeting food. However, Sue Fredericks had deadlines to meet. The O'Haras had arranged several exploratory excursions in and around the ranch, and Sue participated in all of them. When we returned to the house, though, Sue would disappear with an assortment of proof sheets.

Tied at the bottom of the list, in terms of interest to me, were Marty Irwin and Sheila Cooke. Carl and Peggy ran a successful import-export business, and Marty was the business manager and Sheila the head of purchasing. The business was headquartered in Santa Barbara, but Marty was the type of person who looked like he'd never been outdoors in his life—pallid complexion, with bags under his eyes. You see plenty of people who look like that in big cities, but it was sort of a novelty to see one in Santa Barbara. Even though Sheila looked like she spent more time outdoors than Marty, it was clear that the two of them were soulmates—both were dedicated to business and rarely talked of anything but business. The same could have been said of Myron Wallace, but maybe I just found him more interesting because haute cuisine was more interesting than import-export.

It started out as a relaxed and pleasant weekend. Saturday we all rose about eight or nine, and Carl and Peggy served up a lavish breakfast. We packed a picnic lunch and hiked up into the mountains and returned in time for a ranch-style dinner of steak, potatoes, and salad. After dinner, we went into the living room for coffee and conversation. Either the mountains blocked TV reception—even if you had a dish—or Carl and Peggy simply didn't believe in TV because the house didn't have one. As a result, the long-lost art of after-dinner conversation made a comeback Saturday night. A little before midnight, Carl and Peggy appeared with a delicious hot-cider concoction. It can get awfully chilly in the California mountains. About ten or fifteen minutes later, I felt extremely tired and decided to get some sleep.

I'm basically a city boy, but I've always heard that the country air is supposed to make you feel more invigorated. Maybe that's true if you are used to it. I awoke, still sleepy, at ten o'clock Sunday morning, an hour or so later than I usually get up. Actually, I was closer to groggy than to sleepy. It required an act of incredible willpower to get myself shaved and dressed, but the battle between hunger and sleep always ends—for me—in a victory for hunger. I figured that I'd better take care of both of the above items pronto because breakfast had been served about nine o'clock the previous morning, and if I didn't get downstairs soon, there probably wouldn't be any left.

On arriving downstairs, I discovered that I needn't have worried. Even though it wasn't the night before Christmas, not a creature was stirring, not even a mouse. Well, the mice might have been stirring, but the humans certainly weren't. I appeared to be the only one conscious. The others drifted down in ones and twos during the next hour or so. The O'Haras were the last to show up, but nonetheless managed to prepare omelets, Belgian waffles, country sausage, and popovers. I guess when you're a good cook, you can practically do it in your sleep.

I chalked up the overall languor to the mountain air and a surfeit of fine food. Casual conversation during the weekend had disclosed that everyone but Ann lived somewhere along the California coast. The sea air must have a different effect than the mountain air because everyone else noted how well they slept.

We decided that the aforementioned omelets, waffles, sausage, and popovers constituted brunch rather than breakfast and decided to take a final hike into the neighboring mountains during the afternoon. Ann's wardrobe would have been better suited to Cannes than a California ranch, especially as she had forgotten to bring hiking shoes, but Peggy had loaned her a pair that fit reasonably well. We returned early in the evening. Since I had a business appointment early Monday morning, I said good-bye to everyone, thanked Carl and Peggy for a pleasant weekend, and drove directly home. I dropped off my rented car, availing myself of the computerized drop-off arrangement where you just leave the car and they e-mail you the bill. I did, however, write down the mileage on the odometer just to make sure that I wasn't getting ripped off. After that, I removed the small suitcase I had used to pack my gear and walked the couple of blocks home.

When I got to the guesthouse, the message light on the phone was blinking continually, and there was a note from Pete on the door. The note said "Urgent—see me the moment you get back!!" I decided to pick up my messages later and went over to the main house.

"What's so urgent?" I asked.

"Carl and Peggy O'Hara have been calling the house every half hour for the past two hours. They've left messages on your machine. They want you to call back immediately, if not sooner."

Maybe I was such a good guest that they wanted to make sure that I hadn't made plans for next weekend yet. I managed to get through with no difficulty whatsoever.

It was, to say the least, an extremely interesting phone call. I was astounded to learn that Ann Robinson had discovered that she was missing approximately $400,000 worth of jewels and that I, as well as everyone else in the house, was under suspicion of robbery! The moment the theft was discovered, Carl and Peggy had called the police. A thorough search of the house, everyone in it, and their possessions had been made, but no jewelry had been found. I was the only person who had left, but the highway patrol had been unable to stop me because no one had any specifics on the car that I was driving—and not having my cell phone, I couldn't be reached. It turned out that some of the messages on my answering machine were from Carl and Peggy, nearly frantic with worry, and one or two were from the police, inviting me in for questioning.

I sat back and digested this. I knew I wasn't guilty, but I had to admit that I was clearly the leading suspect. After all, no one else had left the scene of the crime.

Speaking of crime, being a detective is a little like being a surgeon. There is a rumor to the effect that surgeons should never operate on someone they love because it clouds their judgment. Well, my judgment was pretty clouded at the moment, so I decided to ask Pete to take a look at it. It took me about an hour to describe the events of the weekend. Pete asked me to clarify a few points and then asked me if I had any ideas.

As I said, I wasn't thinking too clearly. "One possibility is that Ann Robinson is committing insurance fraud. But I guess that's pretty unlikely. We all saw the jewels, and the police searched the house and everyone's possessions thoroughly. They wouldn't exclude the possibility of insurance fraud. Besides, Ann Robinson seemed pretty well fixed. She's some sort of heiress, and if you can bring $400,000 worth of jewelry for a couple of weeks' vacation, you're probably not hurting."

Pete nodded. "As you were telling the story, a couple of things occurred to me. Didn't it strike you as a coincidence that everybody slept late on Sunday, and almost all of them mentioned how sleepy they were?"

I could have kicked myself. "Of course! You think that maybe we were all given some sort of knockout drops?"

"It seems a reasonable possibility. After all, you mentioned that, after you drank the hot cider, you suddenly felt very sleepy. Hot cider would undoubtedly mask any odd taste. Besides, you mentioned that everyone drank it. Carl and Peggy made it, but how was it served?"

I thought back. "They brought it out in a large punch bowl and threw a few logs on the fire. Then they served a cup to everyone. After that, the punch bowl just stood on the table, and everyone helped themselves. I guess that puts Carl and Peggy under added suspicion."

"Yeah, but anyone could have done it."

"I guess so. It looks like everyone had opportunity and means, and only Ann didn't have a motive."

Pete thought for a few minutes. While he was thinking, I was worrying, mostly about my future, and hoping that inspiration would strike.

All of a sudden, Pete said, "I think I've got an idea. Where are the jewels?"

"I don't have them."

"That wasn't an accusation; it was a question. I know you don't have them, but where are they?"

"Not at the ranch house, that's for sure. Could they be hidden somewhere else on the ranch?"

"I don't think that's too likely," Pete said. "The ranch is isolated, so the thief is very unlikely to get back there, and whoever comes back there will immediately become the prime suspect. No, I think it's likely that the jewels are gone."

"Okay, I'll give you that. But gone where?"

"I think the thief took them home with him. Or her."

"But I'm the only one who's home! Other than Carl and Peggy. I don't think they did it, and I know for sure I didn't."

"Look," said Pete, "let's think about the situation from the thief's point of view. He gets there, he sees the jewels, he gets the idea to steal them. He couldn't have known about them in advance because no one knew Ann was coming until the last moment. Therefore, he couldn't have made plans to steal the jewels in advance. Obviously, he couldn't steal them, put them in a package and mail them to himself or to a rented postbox—he just wouldn't have had the materials or the stamps in advance. He couldn't keep them because if the theft was discovered early, everyone and everything would be searched. No, the safest thing would be to steal them, drive home, put them safely away, and then drive back to the ranch before anyone woke up. To make sure no one woke up, he, or she, got some sleeping pills from somewhere and put them in the cider."

All of a sudden, I saw where he was headed. "I think I'm following you, Pete. We need to look at the car rental receipt, right?"

"Yeah."

"They're e-mailing it to me. Let me look at it on my desktop. I never seem to be able to get this cell phone to download attachments properly." Five minutes later, I came back with the relevant information.

When I got back, Pete said, "There were only two cars out there—yours, and Carl's and Peggy's. Didn't you tell me their car badly needed a tune-up, and you could hear it a mile away?"

"He couldn't risk taking that car and have it wake up someone. So he must have taken the rental car."

Pete handed me a slip of paper, on which he had written

$$C = \$60.00 + \$0.15M$$

I didn't have to remember a whole lot of algebra to remember what it meant—the cost C of the rental car was $60.00 plus the number of miles M I had driven times 15 cents a mile. He took a moment to subtract the tax off the bill.

"The rental car cost was $120.30. That means we have to solve the equation

$$\$120.30 = \$60.00 + \$0.15M$$

"The mileage cost comes to $60.30," he continued, "so the car must have been driven 402 miles."

(Algebra continued on p. 162)

"It's about 120 miles from here to the ranch," I said. "That means that I drove the car about 240 miles. So that leaves 160 miles unaccounted for."

"Assuming that the thief drove as directly as possible, it would be around 80 miles each way. That looks like it rules out Marty and Sheila—they both live in Santa Barbara. I guess it's narrowed down to Sue and Myron."

"I'll go in and talk to the police right away," I said. Then I stopped. "Can you give me a lift? My car's still in the shop."

We conveyed our suspicions to the local police. They listened to my story but still told me not to leave town. I could understand their reluctance to abandon me as a suspect, and I spent the next few days in agonized uncertainty. However, I received a welcome phone call on Wednesday from Carl and Peggy. The police had kept a discreet eye on Myron Wallace's home, about 75 miles away to the north (I guess he made junkets into L.A. to check out the restaurants). Myron had an unsavory visitor, a local underworld fence, late Tuesday night, and Ann's jewelry had been found in the fence's possession. I didn't think that dabbling in stolen jewels invalidated his restaurant recommendations, and I still intended to try the ones whose names he had given me.

In a burst of generosity, the police had given credit where credit was due. Ann Robinson had called us full of thanks, and more tangible offers of gratitude. Any reasonable fee and expenses would be immediately paid.

Which led to the question of how reasonable the fee should be. It wasn't much work for us, and we settled for $3,000 for the fee. However, it has always been a good policy to make sure that the client pays for all expenses relevant to solving the case. Of course, that meant the $120.30 plus tax for the car rental. Plus the $250 deductible on the insurance policy for repairing my car—after all, if I hadn't been in that accident, Ann might never have gotten her jewelry back.

CHAPTER 6

MESSAGE FROM
A CORPSE

It looked like my move from New York to Los Angeles was working out well. There's something very pleasant about living in a city where the majority of its inhabitants seem to be easygoing, and on any given day the weather is somewhere between tolerable and idyllic, and generally a lot closer to idyllic than tolerable. It never snows, and even though it gets hot, it's not like the humidity-saturated heat of a New York summer day. On the downside, I still wasn't reconciled to separating from Lisa, but it was clear that Los Angeles had an abundance of prospects. It was just a matter of finding one that clicked.

In addition, it was clear that Pete and I got along well. Yes, I wouldn't have minded if he wasn't so obsessed with sports, but then, sports was an all-consuming passion for a large segment of the population. Also, Pete seemed to have a knack for solving problems of the type that never seemed to arise in New York but kept popping up in L.A. At any rate, when I proposed the idea of formally joining forces, he accepted with alacrity. We had complementary talents, and we felt that between us we could probably do a pretty good job of cutting whatever Gordian knot we encountered. So we had some business cards printed up, and we were in business.

It was understood that Pete's talents did not include hustling up potential clients. He's not antisocial, but he doesn't realize that in order to get a few clients, it is necessary to meet a lot of people. That means accepting a lot of invitations. That doesn't bother me, as I like going to parties. And so I found myself at a party in Beverly Hills. I wasn't exactly trolling for clients, but you have to be alert to opportunities.

In this line of work, opportunities don't so much knock as talk, and so you have to get in the habit of being a good listener. I make it a point to try to listen, at least occasionally, when an individual rambles on interminably about events in his or her life. So I did a good job of listening when Alma Steadman, a wealthy widow I met at this party, told me about her problems. These included having to deal with an executive in her husband's organization who was possibly trying to defraud the company, a son who kept fooling around with actresses rather than trying to find a nice girl, and her widowed sister from Vail who had moved into her Beverly Hills mansion and kept trying to steal her (Alma's) boyfriends. After about fifteen minutes, I gave her our business card, smiled sweetly, and went off to try to make other acquaintances.

Pete was out the next day, and so I happened to be on duty when the phone rang. I picked it up.

"Lennox and Carmichael, Investigations." We had decided to put the Lennox first because it sounds better if the two-syllable name comes before the three-syllable name, especially when the accent of the three-syllable name is on the first syllable.

"Good morning." A slightly nervous female voice that I thought I recognized. "Could I speak to Mr. Carmichael?"

"Speaking." Terse makes you sound like a detective.

"Oh, hello. This is Alma Steadman. You may remember we met at the O'Connor party yesterday."

"I do, indeed. What can we do for you, Mrs. Steadman?"

She hesitated. People are always a little nervous at first with a detective. Then she took the plunge. "I believe I've been robbed, Mr. Carmichael."

"Would you like to come talk to us, Mrs. Steadman?"

"Do you suppose you could come here instead?"

"Certainly. What time would be convenient?"

"How about tomorrow at eleven?"

"That's fine. Would you care to give me a preliminary idea of what your problem is?"

"There's about $400,000—I'm sorry, I have to go." She hung up.

Pete arrived home later that afternoon. My conversation with Alma Steadman had been so brief that I was able to repeat it to him practically verbatim.

"You know, Freddy," he remarked immediately, "there was one very unusual thing about that phone conversation."

It was important to let him know he wasn't dealing with someone who just got out of kindergarten. "You mean the way the conversation ended. Obviously someone whom she suspected was involved had come into the room," I finished smugly.

"I mean besides that. She didn't say, 'I've been robbed.' She said 'I believe I've been robbed.' Now, most of the time, when you've been robbed, you know it."

"Maybe she bought a painting that she thought was genuine, and it turned out to be a forgery. Or fake jewelry."

"Maybe." He looked a little dubious. "But why would she have ended the phone conversation so abruptly? Oh well, we'll find out tomorrow." He wandered over to the TV set and started looking at a random baseball game.

No question, Pete may have had the brains to be a detective, but the heart and soul were a little bit lacking. I tried to recall what she had told me about her problems.

George Wilson was her husband's right-hand man, and vice president in charge of finance for her husband. George and Harry Steadman had gone to Vanderbilt together. George worked for some life insurance firm, but Harry had dangled stock options in front of George. Stock options are tremendous incentives, and they had pried George loose from the life insurance firm, which I gathered was located somewhere in the mid-South. I'm a little hazy on states in that neck of the woods because, let's face it, most of what happens in the U.S.A. happens in the *really* big cities, and there aren't many in that area.

Al Steadman was their only son. He had some sort of executive position that allowed him to do a playboy number with the starlets who frequent Beverly Hills and Hollywood. He had actually done more than a playboy number with one named Vicki Ventana. Vicki was one of those actress types who pop up briefly, get a small but continuing role in a TV series, and then find Mr. Right and start a family. Mr. Right had been Al Steadman, at least until they got divorced.

Gwen Turner, Alma Steadman's sister, had accepted Alma's offer to live with her in Beverly Hills. Gwen was displaying an unseemly interest in Vaughn Ellis, Alma's latest beau. This hadn't got her kicked out of the house—yet—but the relationship between the siblings was strained, though not to the point of fraying completely.

I decided to offer these capsule commentaries to Pete if he was interested. He wasn't. I was seething a little but not as much as I might have been because a no-hitter was in progress, and it's hard to pry Pete away from the tube in such situations. I sat stoically through a hitless fifth, sixth, and seventh inning and then departed. I'm not even sure Pete noticed. I just hoped he'd generate some interest after eleven tomorrow, when it might actually matter.

At eleven the next morning, we were greeted at the door by a butler who told us that madam was waiting for us in the den. He escorted us to its door and exited discreetly. We knocked.

Nothing happened. We knocked again. Still no answer. "Mrs. Steadman?" I asked, feeling that I should be the one to speak, as she might recognize my voice. Still no answer.

The door to the room was open a fraction. "I guess we're just supposed to go in," Pete said. We entered a living room with a collection of elegant impressionist art works, rich leather sofas, bookcases, and a beautiful bar. It was, however, the outsized mahogany desk in the center that caught and held our attention. Behind the desk was a sumptuous leather chair. Leaning forward in the chair, her head on the desk, was Alma Steadman. It was not necessary to be a homicide expert to tell that Alma Steadman was dead, for there was a pool of blood large enough to satisfy a school of piranhas and an indentation in her head that had "blunt instrument" written all over it.

I hadn't had a whole lot of experience with death, but I could tell from the gagging sound behind me that Pete had even less experience with death than I did. His eyes wide, his hand covering his mouth, Pete was throwing up.

I took charge. "Don't touch anything!" I snapped at Pete. I took out my cell phone and called the Beverly Hills police. There's an app for that. Within three minutes, a squad car was at the door, and a minute later we were talking to a Lieutenant Brad Gillette.

Lieutenant Gillette was bald, in his late thirties, and extremely efficient. I noted that he could have used the product with which his name was closely associated, for he badly needed a shave. Contrary to what I had always read about the enthusiasm of the members of the homicide squad for private detectives, it didn't take much to convince Lieutenant Gillette to accept our story that we had been called in to consult about a theft.

Shortly thereafter, a medical examiner arrived. He was somewhat older than Gillette, maybe in his early fifties, and was extremely clean-shaven. He was the first to touch the corpse, excluding the murderer.

"What's this?" he said, as he lifted Alma Steadman's blood-soaked head from the table. Ugh.

Alma Steadman had slumped forward on top of a thick 8½ by 11 manila envelope, on which was written "For Lennox & Carmichael." That was written in ink, but it was not the only thing on the envelope. Outlined in red was a large, bloody "V." I thought that, if this got to the press (which everything does nowadays), maybe it wouldn't be the best of publicity for Lennox & Carmichael, but then I realized that there's no publicity like the publicity that surrounds a murder investigation.

The palm of Alma Steadman's right hand was smeared in blood, but the fingers were clean, with the exception of the tip of her index finger, which was covered with blood. Even a seven-year-old could have made the obvious deduction that she had dipped her finger in blood to scrawl that last letter and then expired.

"I believe I'll look at this," Gillette drawled. "It might be evidence." Pete and I conferred briefly and put in a demurrer.

"It's addressed to us," I stated emphatically. "You can't be certain until you open it that it's evidence, and we think we're entitled to look at it while you're opening it. If you like, I can contact a lawyer who would know for sure, but that would almost certainly delay things, and you probably don't want that." Rule 1—always be cooperative with the police, especially if it doesn't cost you anything, but citizens have rights and you don't want to give up those rights, even if you don't always know exactly what they are. That's why we have lawyers.

Besides, it almost certainly was evidence, and then Gillette would undoubtedly hang onto it.

Gillette hesitated and then agreed. He opened it. There was a check to us for $5,000 marked "retainer," and a lot of bank statements relating to the Alma Steadman Trust. There was also a scribbled note on which was written, "Al says 6% per year."

We took a look at the statements. On January 1, 2004, Alma Steadman's account showed a balance of $2 million—a lot more zeroes west of the decimal point than on my bank account. There were a whole lot of deposits and withdrawals, but ten years later, on January 1, 2014, the balance was $3,197,385. This is not chopped liver.

Gillette had an app for that as well and was doing some calculations on his cell phone. "Ten years at 6% per year is 60%. Sixty percent of $2 million is $1,200,000. The account should total around $3,200,000, and that's where it is. I don't see why she called you in." He rooted around the statements. "Unless there are a lot of unauthorized withdrawals here."

I took a look at the statements. There were certainly a bunch of withdrawals, but there were many deposits as well. It would take some checking, but I didn't think I'd find any major discrepancies—certainly not $400,000 worth.

All of a sudden it hit me. There were 10 years' worth of monthly statements. Notations on them were made in handwriting that certainly looked the same as the handwriting on the scribbled note. "Unauthorized withdrawals don't figure to be the problem," I stated. "It seems pretty clear that she went over her statements with a fine-toothed comb every month. She'd spot an unauthorized withdrawal in a minute. Incidentally, getting 6% a year in a recessionary interest climate is pretty impressive."

Pete had been strangely (for Pete) silent. All of a sudden he spoke up. "I think I know what happened," he said.

Gillette and I looked at him. "Banks don't pay simple interest; they pay compound interest," Pete explained. I could have kicked myself. Of course! "If the deposits and withdrawals even out, at the end of the first year she would have made 6% on $2 million, which is $120,000, assuming that it was compounded annually. Her balance at the end of the first year should have been $2,120,000. The next year, she would have made 6% on $2,120,000, which is about $127,200. Compound interest means that not only does the amount you deposit, the principal, earn interest, but the interest that you make also earns interest."

"Could it make that large a difference?" Gillette asked.

⌐ (Compound interest continued on p. 168)

Pete nodded. "Ten years of 6% simple interest would be a return of 60%, as you calculated. But I just checked with an app on my smart phone, and 10 years of 6% compounded annually gives a return of about 79%, and even more if it is compounded more frequently. Anyway, that's a difference of 19%, and 19% on $2 million comes to . . ."

"About $400,000!" I said enthusiastically. "That explains why she said that she believed she'd been robbed of about $400,000. All we have to do is find out who arranged for the money to be transferred, and I'd bet dollars to doughnuts it's her son, Al. After all, she mentions that it was Al who said it was 6% annually. Maybe he thought he could sneak the difference between simple and compound interest past her. Maybe he also thought she'd be so delighted with 6% in the current business climate that she wouldn't notice."

Gillette scowled. "Even assuming you're right, there's still that 'V' to be explained. Anything you know about any of the suspects that has a 'V' in it?"

Well, I now had a bigger audience for my capsule commentaries regarding the major players in the drama. I went over them, with particular emphasis on the letter "V."

"There's George Wilson, who was the vice president for finance in her husband's corporation. There's a 'V' in vice president." Gillette scowled again. He had a face made for scowling. Pete was similarly unimpressed. "He and Alma's husband both went to Vanderbilt." Both of them scowled.

"What's his motive?"

"Alma seemed to think he was messing with the books of the corporation. Anyway, next on the agenda is son Al." I thought for a moment. "Oh, yeah! He was married to Vicki Ventana! Remember her?"

Gillette nodded. As expected, Pete drew a blank. Gillette thought for a second and said, "Could Vicki Ventana have been involved?"

I shrugged. "You've got me. Check her alibi. Alma's son and sister live with her." Gillette looked at her appointment book and pointed out that Wilson had a ten o'clock appointment.

When Gillette got off the phone, I continued, "The last member of the household is Alma Steadman's sister, Gwen Turner.

"What about 'V'? She moved here from Vail." More scowls. "She was interested in Vaughn Ellis, Alma's boyfriend." At least I had stopped the scowling, for they both looked thoughtful.

Pete had stayed silent throughout the duration, content to let the real detectives (Gillette and yours truly) work on the problem. He had spent his time staring at the envelope with the bloody "V." But Pete evidently had an idea because he suddenly entered the conversation.

"I think I've got it!" he exclaimed, looking pleased.

I'd seen that look before, but Gillette hadn't. I'd also learned that Pete was very ego-involved in his solutions to puzzles. That can be a good thing, but it can also cause a solution to be blurted out before money has changed hands. I'd have to caution him about this in the future.

However, I didn't see any way that the solution to the murder of Alma Steadman would be worth anything to us financially. I should add that, as a detective, I was well aware of the adage that the person (or persons) who discover the body automatically moves to near the top of the most-favored suspect list. It wouldn't have surprised me in the least to learn that Gillette was having us checked out at this very moment. So I let Pete continue.

"I think it's a dying clue," Pete stated. "I think Alma Steadman was trying to name her murderer."

Gillette looked disgusted. Even I looked disgusted. "We've been working along those lines," Gillette said dryly. Gillette and I had one thing in common; we believed that people who speak slowly do so because they think slowly. Well, I used to believe that before I met Pete. Gillette, however, didn't know Pete and looked at me with an unspoken question: Is your buddy a little dim between the ears?

"Then why wouldn't she just write the name of her killer?" Pete asked. "I'm sure that's what I'd do."

"Maybe she thought the killer was watching, and she wanted to leave a clue that the killer wouldn't be able to decipher," I suggested.

"I don't think so," Gillette interjected. "The first thought that a murderer usually has is to get out fast. According to your story, the murderer was apparently in such a hurry that he or she forgot to close the door. Besides, according to the doc, Mrs. Steadman died shortly after she was hit over the head."

"I'm glad to hear that," Pete said. "It fits in perfectly with my theory."

"Which is?" I prompted.

"That Alma Steadman named her killer."

"But there are two people whose names begin with V," I remarked. "Her boyfriend and her former daughter-in-law. And they're both long shots. Her death seems to be connected with the information in that envelope, and neither Ventana nor Ellis had anything to do with Steadman back in 2004."

"Look," Pete said, "if the killer was Al Steadman, she would have started out by writing an 'A' for Al or an 'S' for Steadman or son. So I think he's eliminated."

"If you're right, that leaves her sister Gwen and George Wilson," Gillette stated. "But which one?"

"Alma Steadman knows she is fatally wounded," Pete replied. "She hasn't got much strength left. Maybe she senses that, if she tries to write the first name of her killer, she won't make it past the first letter. Remember, their first names are Gwen and George, and both start with "G." So she starts to write down the last name of her killer."

"But neither Turner nor Wilson starts with 'V,'" Gillette observed.

"I think she died halfway through the 'W' of Wilson."

Gillette and I looked at each other. "Could be," he remarked, "but I'll have to check it out."

The case was wrapped up in less than twenty-four hours. George Wilson, who had been the executor of her husband's estate and in desperate need of funds to support a failing real estate venture—of which there were lots after the crash of 2008—had indeed been looting Steadman's business. Alma had gotten wind of it and had confronted Wilson with the evidence. Wilson had somehow managed to get behind her on some pretense, grabbed a heavy marble ashtray, and clobbered her. They found a thumbprint with traces of blood on the door.

According to Gillette, we were free to go. He said he'd have to keep the retainer check until they closed the case. Knowing that you can't cash a check written by someone who has died, I had no objection, but it always hurts to bid farewell to $5,000.

A few days ago, I received a call. It was from Brad Gillette. He told us that everything had been cleared up. It was true that Al Steadman had stolen $400,000 from his mother's trust fund, but since he inherited far more than that, there was no one around to prosecute and no point in doing so anyway.

"By the way," Gillette said, "you should be receiving a check for $5,000 within a day or so."

I nearly dropped the phone. "From who?"

"Gwen Turner. I told her about your partner's deductions and suggested that you deserved to be paid. This isn't for publication, of course."

I was still stunned. "That's very decent of you, Lieutenant."

"Captain. I was coming up for promotion, and they were so pleased with the way I handled the case that they speeded it up. I may have omitted to mention that, when I filed my report, I took credit for some of your friend's deductions." He hesitated. "Of course, this isn't for publication either."

"Of course," I agreed. When I told Pete, he agreed that $5,000 in the bank and the friendship of a captain of the Beverly Hills police was well worth not taking credit. So if you run into a Beverly Hills police captain, late thirties, bald, and needing a shave, I'd appreciate it if you kept quiet, too.

Author's note: The notes for this chapter are on the mathematics of finance, which is probably not the world's most interesting material—until it is your own money that is on the line. Even though you may not want to read this now, it wouldn't hurt to keep it handy for when you make a major purchase, such as a car or a house. There's so much money at stake in these purchases that it's worth spending a little time understanding the math, so you can make the deal that works best for you.

CHAPTER 7

ANIMAL PASSIONS

Pete disconsolately pushed aside his plate, leaving the remains of the tofu salad, the day's featured offering at our neighborhood vegetarian restaurant, unconsumed. I followed suit.

"Nobody said it would be easy, Pete," I remarked as we got up to leave. "But then again, nobody said you had to join me in this vegetarian experiment. I must admit, though, that I'm glad you did. Misery loves company."

What, you may ask, were Pete and I, red-blooded (and red-meated) Americans doing at an eatery specializing in vegetables? We both had excuses—there has to be an excuse for deviating from nature's plan. In case you need a refresher course in nature's plan, more than three billion years of evolution have gone into producing *Homo sapiens*, an omnivorous species. This means that we are able to eat everything, and nature intended us to eat everything. But every so often, things get a little out of hand. Pete's excuse for eschewing animal protein was that his cholesterol level had gone off the chart. But I had started this whole thing.

My excuse for a meatless diet? Let me take you back in time to a message on my cell phone roughly ten days ago.

"Freddy? It's Lisa. I'm coming out to L.A. for a visit. Can I stay with you? Give me a call. Miss you. Bye."

Lisa hadn't said she missed me for months. I dialed one of the few numbers I never have to look up.

"'Lo?" Lisa always answers the phone so that the first syllable of "hello" is clipped off.

"Hi, Lisa. Got your message. Just tell me when, and I'll clear my calendar."

"Well, I'm arriving a week from Monday and leaving the Friday after that."

I calculated rapidly. Two work weeks (Monday through Friday), one weekend came to 12 days, minus parts of two days. We could mend a lot of fences during that period.

One of the things that I always found endearing about Lisa was that she had managed to retain a refreshing 1960s idealism in the second decade of the twenty-first century. If it seemed like a worthwhile cause to her, she joined it and could usually be found marching in the vanguard. Currently, the vanguard in which she was marching was animal rights. Anyway, animal rights activists from far and wide were about to drop in on Los Angeles for a national conference. How would I feel about putting her up for the duration of the conference?

As you may have gathered, I hoped it would be a long conference. I hung up the phone, feeling better about things in general than I had in many months.

A couple of weeks later, I met Lisa at the airport. I hadn't realized how much I missed her. Well, maybe I had realized. I looked into those incredible clear blue-green eyes and realized that I was still carrying a torch and probably always would.

While we waited for her luggage, she asked me if I would accompany her to a party she had to attend before we went back to Brentwood. Her wish was my command, so we headed up through the hills of Bel Air to a gated mansion behind which lived a noted philanthropist who was devoting the latter years of his life to undoing his robber baron image. My car was parked by a valet, who curled his lip slightly at being forced to drive something other than a Rolls or a Mercedes, and we headed in.

Food and drink were present in abundance. I liberated a couple of glasses of champagne from the tray of a passing waiter, and we toasted the success of Lisa's visit. Not only was I thirsty, but I was also ravenous. A passing platter of paté beckoned. Either animal rights did not extend to edible animals, or the philanthropist had just off-loaded party arrangements on a local caterer and didn't worry about the details. Also, I reflected, the conference probably entailed more than just preaching to the chorus. At any rate, I was moving in the direction of the beckoning paté when someone else spread some on a cracker. Lisa shuddered.

"Do you realize," she said, looking up at me with those clear blue-green eyes, "that over 25 geese must be killed in order to produce a single pound of foie gras?"

As you have no doubt divined, I was about to make a move to grab some of the aforementioned paté for myself when a sixth sense warned me to divert my attention to a stick of celery instead. I only batted one for two, however, as my sixth sense did not prevent me from replying, "No wonder it's so expensive."

"Strasbourg geese," Lisa continued, "are confined and forcibly fattened in order to produce overly large livers. But if you think that's bad, do you know how veal is produced? It makes me ill just to think of it."

Not me. I salivate like Pavlov's dogs at the thought of veal piccata. But there is a time and a place for veal piccata, and this clearly wasn't it.

"Revolting," I agreed, looking longingly at a plate of langoustine passing just out of reach, and deciding that the denizens of the sea were probably fodder non grata as well. Good thing I made sure the fish tanks were clean before leaving home. "It's times like this that I'm glad I've become a vegetarian."

Well, I thought to myself, as the aroma from a delicately spiced sausage drifted tantalizingly up to my nose, it's not a complete lie. There have been periods when I eat no meat. These periods, however, tend to coincide with the periods between meals.

Like any good detective, I had picked up a clue and acted upon it. Lisa grabbed my arm, looked up at me with blue-green eyes that shone as they had during those good times when we always seemed to be on the same wavelength, and declared, "What a coincidence! So have I."

I soon discovered that Lisa had not only become a vegetarian but she was also a member of the shock troops leading the battle for animal rights. It did not take her long to enlist me in the movement.

In retrospect, maybe I should have resisted. But Lisa can be very persuasive. At least, she never seems to have any difficulty persuading me. Less than an hour after seeing her again, I had converted to vegetarianism. By the next day, she had enrolled me in an animal rights group.

When I told Pete that I was on a meatless kick, I expected a strong adverse reaction. After all, we're talking about a man who never met a cheeseburger he didn't like. What I got instead was a mild, thoughtful one. Pete mentioned that his cholesterol count was way too high and said that if he went on a meatless diet for a while, this situation might be rectified. I certainly wasn't going to argue, especially since I might have had second thoughts if I saw Pete chowing down on thick New York cuts while I was doing rabbit imitations among the alfalfa sprouts.

So now you are up to date. The bank account was in good shape, and so I was well placed to devote my full attention to Lisa. This meant, to Lisa at least, that I could be co-opted into helping with Lisa's animal rights organization. I certainly didn't mind. Things were going unexpectedly well between the two of us. Maybe we just needed some time away from each other, or maybe ours was a love that thrived better in the warm climate of laid-back southern California than in the frigid winters, steamy summers, and disposition-frazzling hostility of New York.

Not only was I taking care of a lot of organizational details for the animal rights crowd, I was also showing up at some of their functions. In exchange for my help, I had become a card-carrying member of an organization whose name I hadn't even bothered to memorize but whose card I had stuffed into my wallet. Oh well, it made Lisa happier.

The aforesaid organization had two upcoming activities, a lobbying effort for an animal rights bill now under consideration by the California legislature and a midnight raid on a university animal research lab. Lisa had asked me to do a survey of 43 members to find out which, if either, they wished to attend. I checked; 8 had been unavailable for either, 33 had been willing to lobby the legislators, and 12 had demonstrated interest in the covert raid on the animal laboratory. I had volunteered for the lobbying effort, but Lisa, who was naturally part of the hard core who were willing to do both, talked me into joining the midnight raiders as well. She asked me how many people were, like her, willing to do both, and I had to admit that I didn't know. I presented the problem to Pete the next time I saw him.

The next time I saw him turned out to be by the refrigerator in the kitchen in the main house. He had the door open and was casting longing glances at a couple of pounds of chopped round. I shut the door of the refrigerator gently and diverted him with a can of Coors and a bag of potato chips. When his appetite had been partially assuaged, if not his longing for animal protein, I asked him how many people in Lisa's organization were willing to participate in both activities.

His eyes closed for a moment, possibly to prevent tears from coming to them as he looked at the hamburger, and then he answered, "Ten."

Since Lisa was probably going to demand some sort of accounting, I asked Pete how he came up with that number. Just in time, too, for as his eyes opened, a faraway look came into them, which I interpreted as an urge to barbecue, and then consume, part of a cow—or possibly all of it. I won't say I blocked the refrigerator door, but I positioned myself between it and Pete.

He could tell what I was doing, but he had not yet reached the "Out of my way, Freddy, or I'll barbecue you as well" stage. He sat down and said, "Of the 43 members, 8 were unable or unwilling to participate in either activity, leaving 35. If one adds the number of people who are going to lobby to the number of midnight raiders, the total is 33 plus 12, or 45. However, those willing to do both have been counted twice, once for the lobbying effort and once for the laboratory raid. Since 45 less 35 is 10, 10 people must have been counted twice. I take it that Lisa has been counted twice."

⟋ (Fundamental Counting Principle continued on p. 182)
"You take it correctly, Pete. But I have also been counted twice. Lisa has talked me into doing both."

"Freddy, I can see why you find her attractive. But she's an artist, and artists often have very strange ideas. Have you stopped to consider that interfering with laboratory research might have unfortunate consequences for those whom the research might benefit? Plus, let me add, that there are laws prohibiting certain types of actions in this regard. I don't know the specifics, but a few years ago members of some extreme animal rights organization set fire to a UCLA researcher's car. Or something like that."

"I'm not planning on committing arson. I did mention the laboratory research aspect. She says the ends do not justify the means. I let her have the last word in the interests of harmony."

"The balance between desirable ends and necessary means is indeed one of the oldest of philosophical dilemmas." Having elevated the tone of the discussion substantially, Pete marched out of the kitchen, possibly hoping that I too would disappear, so that he could sneak back and make mincemeat of the hamburger. If I knew him, he would then go to the grocery store and replace the chopped meat to cover up the crime.

The lobbying effort took place Friday afternoon, and the laboratory raid was scheduled for Friday midnight. The former went without incident. After a romantic candlelight dinner, Lisa and I prepared ourselves for the midnight raid. Dinner was so pleasant that we barely made it in time. In retrospect, I wish we hadn't. You can imagine our surprise when, after penetrating the laboratory's defenses, which were pathetically easy to surmount, we ran into the police! It was difficult to deny our intentions, especially when we entered carrying cages in which we intended to exhibit the animals we had planned on freeing at a press conference to be called the next morning. As a result, we almost spent the night as the guests of the city of Los Angeles, but we were released on our own recognizance in the wee hours of the morning. It does not look good for a practicing detective such as myself to have an incident like this on his record.

I woke up late, to find that Lisa had already departed. A note told me that she had proceeded to conference headquarters. I encountered her, bright-eyed and bushy-tailed, in the lobby.

"Oh, Freddy, isn't it just wonderful?" Even after a late night, those blue-green eyes were still clear. Mine were neither blue-green nor clear. I was just barely functioning.

I'm not so good-humored when I'm short of sleep, and since I only had five hours of sleep, I was somewhere between grumpy and totally bitter. "What's so wonderful?" I snapped.

For reply, she handed me a copy of the morning paper. We had made page three of the metro section, complete with picture. Lisa, I was not surprised to discover, was remarkably photogenic. Mercifully, neither my photograph or name appeared in the article. I silently gave thanks for small favors.

"Look, Freddy, at all this wonderful publicity. Do you realize that as a result of last night's incident and the publicity we received, that our membership has increased by thirty-eight?"

In the back of my brain, without my knowing it, a decision had been coalescing out of the inky blackness.

"Thirty-seven," I said.

Lisa look puzzled. She counted some registration cards in her hand. "No," she said, "there are thirty-eight registration cards here. And I'm sure we'll get more."

"I'm sure you will," I said, "but as of this moment the number is thirty-seven." I removed the membership card from my wallet and tendered it to her. "I resign." Rarely have I been handed such an opportunity for such a boffo exit line.

I returned home and told Pete of the day's events. He nodded sympathetically.

"It's probably for the best, Freddy. Lisa certainly seems capable of persuading you to exercise poor judgment."

I concurred. "I'm sure I've learned a valuable lesson. It's just going to take me a little time to figure out precisely what that lesson is."

Suddenly, I perceived a golden lining in the clouds that had hitherto blocked the sun. "By the way, Pete, have you any idea what has happened to your cholesterol count?"

"It's down a bunch. Probably so is yours."

"What would you say to a nice, juicy steak for dinner?"

"I think it would go great with some cheeseburgers as appetizers. I'll go the store and get some."

"While you're at it, get some bacon for breakfast. And sausage. Better yet, why don't you come back with about fifty pounds of assorted meat?"

Pete grabbed the keys to the car and left in a hurry, looking happier than I had seen him in days. He returned shortly, and while we were putting the results of his shopping expedition in the refrigerator, a question suddenly popped into my head. I thought about it for a moment and then gave it voice.

"Something strange just occurred to me, Pete."

"Oh, yeah?"

"I wouldn't have thought that Lisa's animal rights group contained any stool pigeons. As events proved, I was the only one not satisfied with the results of the evening. And I certainly didn't squeal. But the fuzz were waiting for us when we arrived at the laboratory. They must have been tipped off."

Pete beamed. "A flawless piece of deduction, Freddy. I told the cops about your planned raid on the lab."

My jaw dropped. "You *what*? Did you know that I almost spent the night in the West L.A. precinct house?"

He nodded. "I felt that I had to take that risk. If they caught you after your raid had proved successful, the penalties would have been much more severe. In addition, it seemed clear to me that although the goals of the animal righters might well have been laudable, the means they were choosing to pursue them were open to question. Recently passed legislation makes interfering with legitimate experimentation a felony. Intended interference isn't so bad. Don't tell me you wouldn't have minded spending time behind bars."

Although he was probably right, I felt that I had to make some sort of protest.

"Did you consider giving me a voice in all this, Pete? Has the absence of cheeseburgers from your diet affected your thinking processes?"

"*Nothing* affects my thinking processes," Pete snorted. "First of all, I do not consider it my place to interfere in your decisions. [Since when?] Second, if I had discussed these plans with you, the outcome would have been counterproductive. You would have told Lisa, and no good could have come of that."

I had calmed down a little, and I saw what he had in mind. "I guess you're right. And I admit I was under her spell. And still am, for that matter, although a night of incarceration might have made me see things a little differently." I reflected a little further. "By the way, Pete, if I ever get talked into a similar rash move such as not eating meat in the future, do me a favor and don't participate in it with me. When it comes to devising risks for me to take, you simply don't know the meaning of fear."

On a scale of one to one hundred, Lisa's visit rated about an eighty-seven. I told her this as she was about to get on the plane for New York, not mentioning the fact that her rating of the visit would probably plummet drastically if she ever learned of Pete's interference. She laughed.

"That's a lot better than some of the time we spent together when we were married."

"We still are, unless you filed for divorce."

"No, I haven't done that. In fact, . . . ," she checked herself. It didn't require an advanced degree in psychology to know what she was thinking. Then she gave me a great hug (Lisa gives great hugs), kissed me tenderly, told me to stay in touch, and got on the plane.

The macho detective at this stage stops at the airport bar for a quick one and leaves without a second thought. Not me. I kept waving until the plane was out of sight.

CHAPTER 8

NOTHING TO CROW ABOUT

"And don't forget the birdseed!" Pete yelled at me as I left for the supermarket.

Since it was the third time he had reminded me, I certainly wasn't going to forget the birdseed. Lyle Carson, an old friend of Pete from his undergraduate days, was scheduled to cross our doorsill in just a few hours. Some of Pete's friends are rather strange, but none eat birdseed. The birdseed was for Sweetlips, Lyle's pet crow.

Lyle turned out to be a gangly dude in his late twenties who pulled up a few hours later in a late model Porsche—earned, no doubt, from plying his trade. According to Pete, Lyle was a contemporary Cincinnati Kid, a professional poker player with a weakness for birds. Had Lyle not discovered poker, he would probably have become a veterinarian. Or an ornithologist.

Lyle was in town for two reasons. The first was to prepare for the World Series of Poker, held annually in Las Vegas. There are a number of events now in the World Series of Poker, and the biggest one requires an entry fee of $1 million. That was out of Lyle's league, but well within his reach was one that had a $10,000 entry fee and the winner-take-all first prize was over a million.

Lyle was a southern boy who had played in most of the cities of the Deep South and probably a number of the riverboat casinos as well. At one stage, the South was the center of the poker world, but with the advent of televised Texas Hold 'Em and the World Series of Poker, that center had shifted to Las Vegas. However, the casinos in L.A. played Hold 'Em and attracted a lot of the better players, and Lyle wanted to brush up before testing the waters in Las Vegas. I asked him why he didn't go to Las Vegas directly, and he said that Las Vegas got him so wired that he'd never survive the WSOP. Also, there was a little matter on which he wanted our advice.

Since Lyle obviously knew when to hold 'em, and knew when to fold 'em, he wasn't there to ask our advice on poker. The three of us were seated in the living room when Lyle pulled out several copies of a newsletter and handed them to us. I took a look at my copy. The newsletter was about eight pages or so and was titled "A Summary of Computerized Handicapping Systems."

A quick scan convinced me that the newsletter writers had gotten hold of a potentially profitable idea. They evidently subscribed to a number of computerized handicapping services and tracked them, printing individual predictions of each service, as well as an assortment of statistical indicators. The various services were ranked in terms of percentage of winners, percentages of profit or loss, ability to select longshots, etc.

Pete looked at his copy and turned to Lyle. "So you're still playing the horses. Doing any better?"

Lyle winced. "'Fraid not." He paused, and then continued. "They say every professional gambler has a weakness. Some bet on sports, some do drugs, some have messed-up personal lives. With me, it's the ponies."

I looked at the late model Porsche parked in our driveway. "Seems to me you do all right."

Lyle nodded. "But I'd do a lot better if I could beat the horses, or at least stay even. And I think I've got a chance."

Pete frowned. "By subscribing to," he scanned the list, "the Brooklyn Handicapping System?"[1] I looked at my list and saw that the boys from Flatbush were currently ranked highest.

I (For more information on terms in this chapter, see An Introduction to Sports Betting on pp. 231–33 and Notes to Chapter 8 on p. 235.)

Lyle shook his head. "Naw, the rankings change frequently. They're just hot now because they picked a few longshots[2] at Aqueduct. Let me show you three letters that were included with my last few issues."

The first letter Lyle showed us was printed on good-quality stock, and the first portion of the letter said that Mr. Carson (the name appeared in the body of the letter, personalizing it through the magic of computerized mailing) had been selected to participate in a revolutionary advance in computerized handicapping. There followed a paragraph that included phrases such as "neural nets" and "massively parallel processing." It meant nothing to me, and I skipped to the second paragraph.

I had no difficulty following the second paragraph. It predicted that a horse called Second Opinion would win the third race at Belmont on July 22.

Pete had evidently arrived at that point in the letter as well, for he stated, "You wouldn't be asking me my advice if Second Opinion had lost, so obviously it won."

"Breezing," Lyle agreed. "By four lengths on a muddy track. But the horse was the class of the field, and went off as a 5 to 2 favorite in a six-horse race. So I didn't really think much about it until the second letter arrived."

We both took a look at the second letter. After a little bit of self-congratulation concerning the victory of Second Opinion, the letter suggested that the recipient bet on a horse appropriately named Risky Business, which was to run in the fifth race at Belmont on July 30. Combining Lyle's visit with 20–20 hindsight, I of course realized that an investment in Risky Business had proved to be anything but.

Lyle had gone to the kitchen to get a beer. Even though he was in his late twenties, I could see that he probably got carded in bars. When he came back, he said that Risky Business had finished first in a field of eight by a couple of lengths on a fast track. He had played a hunch and bet a hundred bucks through a bookie. Risky Business paid a little more than 6 to 1 to win, so Lyle found himself better than $500 to the good.

"By now I was looking forward to the next letter. Admittedly, a favorite and a 6-to-1 shot doesn't qualify you for Handicapper of the Year, but I get a lot of this type of junk in the mail, and most of them tout losers. Anyway, I had kind of expected a pitch by now, but the third letter was more of the same. I was a little surprised to see that it was selecting a small stakes race on a northern California track, but I went for it, especially as I was playing with house money. I bet five hundred to win on Golden Destiny. I didn't really expect it to happen, but Golden Destiny ran away from a field of ten, and I am now better than $10,000 to the good. Golden Destiny was a little more than a 20-to-1 longshot!"

Pete whistled. "Lyle, you're ten grand to the good. What do you need me for? Just continue playing on house money if they send you tips. And if they ask for money, don't go into your own pocket."

Lyle stretched his legs and scratched the jug ears. "Yeah, Pete, I've played enough poker with you that I *knew* that's what you'd say. You never did like to reach into your jeans for more dough. And I would have done just that, except that there's a wrinkle to it."

With that he handed us a fourth letter and a prospectus for a software development corporation. The letter recapitulated the successes of the previous three bets and mentioned that the programmers and AI (artificial intelligence) specialists who had developed it were planning on improving the system. They were forming a corporation using venture capital. Investment units would be $100,000, so the number of owners would be small. Plans were to use a percentage of the initial capital for development and a percentage for wagering. With money gained from betting, the corporation would develop software for stock, commodity, and currency speculation. It was hoped that the corporation could go public in five years. The prospectus cited past rewards to investors in such diverse areas as computers and bioengineering.

Pete studied the prospectus and the letters for a few minutes, and then he turned to me. "You're the expert in this area, Freddy. What do you think?"

I gave it some thought. "It might be on the up-and-up. Of course, what they say is mostly technobabble, but if they really had something, they wouldn't give away their hole card.[3] I've seen lots of similar prospectuses. Most end up going under, but then most ventures go under. However, if someone had handed me the prospectuses for Apple or Genentech, I'd have made the same comment. One thing for sure, they're batting three for three."

Lyle digested this and washed it down with a swig of beer. "And what do you think, Pete?"

Pete took a moment to reply. Finally he said, "There may be something in what Freddy says. But I think it's just a slick version of the old Chinese Restaurant Principle scam. Pocket the $10,000 and call it a day."

Lyle was obviously rooting for Pete to go the other way, for his disappointment showed clearly on his face. Maybe he only put on his poker face at poker. "What makes you say that, Pete?"

Pete tapped the letters. "I can't be sure, but here's what I think happened. The newsletter has over a half million in circulation. There were six horses in the first race, eight horses in the second, and ten in the third. That means there are $6 \times 8 \times 10$, or 480, possible ways of selecting the winners of all three races."

Lyle was not up with Pete. "Where do you get that number?"

Pete thought for a moment. "It's like a daily double.[4] If there are six horses in the first race and you pair each of those horses with all eight horses in the second race, you'll get six times eight daily double tickets. That's forty-eight tickets. Now you have to wheel[5] each of those forty-eight daily double tickets with each of the ten horses in the third race. That's 48 times 10, or 480."

Pete had obviously chosen the right way of explaining it to Lyle, for Lyle said, "I think I see what you're trying to say. They just sent out a whole lot of letters. If they sent out half a million letters originally, they would have more than 1,000 people who were given all three winning horses. And the fact that a longshot came through gives them a lot more credibility."

Pete nodded. "I'll give them credit for dressing it up with that venture capital stuff. Years ago, they just would have asked you to ante up for their picks in races."

"I guess I'll take the money and run. Or just use it to pay my entry fee into the World Series. Speaking of fees, what's yours? You've just saved me a bundle."

Pete shook his head vehemently. "If it's a legitimate deal, I may have cost you a fortune. Besides, you're an old buddy. No charge." I wasn't so wild about Pete giving away advice, our stock in trade, for free, but it was his advice and his friend.

"Can't let you do that, Pete." Good for Lyle! Maybe there was hope for a commission yet. Lyle frowned, and then his face lit up. "How about this? If I'm the big winner, you get 10% of $1,000,000. If I win a million, I'm not going to begrudge a mere hundred grand." It may be mere to you, Lyle, but it's not to me. Half of a hundred grand is fifty grand. I could go for that. I didn't think that Pete would object to fifty grand, either.

And thus it came to pass that we were minority shareholders in Lyle Carson. I'd have to have a talk with him about keeping a poker face before he left.

Lyle's practice sessions were generally two- or three-day binges in the L.A. casinos, during which time we were entrusted with the care of Sweetlips. Lyle had departed for a poker palace in Gardena one evening, and it was Pete's turn to feed Sweetlips. Pete had opened the door to Sweetlips's cage when the phone rang. Pete's attention was diverted for a moment, and that's all it took. Sweetlips was out of the cage and through an open window faster than the proverbial bat out of you-know-where.

Pete made a few choice comments, none printable. "Lyle will kill me! Sweetlips is his lucky crow!"

I couldn't believe my ears. "Lucky crow?"

"Lyle's superstitious. Lots of gamblers are. Ever since he got Sweetlips, he's been on one humongous winning streak." Pete raced out into the yard. There, high on a branch of the tallest pine tree in the front yard, Sweetlips was savoring his (or her) freedom. Desperately, Pete waved birdseed at Sweetlips. Sweetlips regarded it with contempt.

Visions of Lyle playing abysmally at the World Series danced through both our heads. As might have been expected, Pete was there first with the obvious solution. "Freddy, we've got to come up with a substitute crow!"

Of course. What else? We tried about ten pet shops. No crows. We looked through the *Recycler*. Parakeets, budgies, canaries, and cockatoos. But no crows.

Things were getting desperate. Then, just as things looked black (an appropriate color, considering the missing party), I came through. I recalled that a casting director acquaintance of mine had told me about someone who supplied birds to movies. A twenty pried the name and number out of her. The birdman of West Covina was in, and he had a choice of three crows we could buy. We piled into Pete's car, and three-quarters of an hour later we were ready to purchase a crow.

Pete looked at me. "Freddy, you spent more time with Sweetlips than I did. Which of these three birds looks the most like him?" Pete didn't seem troubled by the he-or-she problem.

When you've seen one medium-sized crow, you've seen them all. It was a case of eenie-meenie-minie-crow. I voted for the one in the middle.

We turned to the trainer. "How much?"

"That's $750."

Whoever said that a bird in the hand was worth two in the bush was obviously unfamiliar with current market conditions. "Crows are everywhere. Why so much?" I asked.

"They're hard to catch." Tell me about it. "Besides, they rent for $200 a day if the movies want them."

I turned to Pete. "I'll use the company credit card. Maybe we can claim a deduction."

Luckily, we arrived home with Sweetlips II long before Lyle showed up. Evidently Lyle was fooled by the deception, for he departed for Las Vegas several days later with Sweetlips II in tow, not having the slightest inkling that he was no longer in possession of his lucky crow.

As it turned out, we were going to have a front row seat to the World Series of Poker without leaving home. As usual, ESPN had opted to telecast the entire World Series of Poker live. With a potential $100,000 or more on the line, we were going to be glued to the screen. At least, I was. Two sleepless days of TV-watching later, they had gotten rid of the amateurs, and there was Lyle at the final table!

Pete had been smart enough to catch up on sleep during the prelims, whereas I, like a dummy, had a bad case of strained eyeballs. So Pete was bright-eyed and bushy-tailed as the field narrowed to four. If it was hard for me to sleep, I wondered how it was for Lyle. He looked groggy, but there was more than $800,000 in chips in front of him when the fourth day's play started.

If this had been Hollywood, there would have been lots of drama building up to a climactic hand. There wasn't. Maybe Lyle had reached his level, maybe he was just exhausted, or maybe his luck was simply bad. Anyway, the money just started dribbling away. Five hours later, he shoved in his remaining chips, about $250,000, on a situation in which Pete later figured he was a 3-to-1 favorite. Unfortunately, his luck ran out. On the last card, his opponent made a flush to Lyle's three queens. Lyle smiled a weary smile, murmured the words that Arnold Schwarzenegger made famous in *Terminator I* ("I'll be back"), and walked out, possibly to trade Sweetlips II for a new lucky crow.

I cursed. Pete looked down at the floor and then slowly raised his face to me. It was clear that he was fighting an internal battle. Finally, for one of the few times since I have known Pete, a suppressed irrationality surfaced. He turned to me and said accusingly, "Freddy, why couldn't you have picked a different crow?"

Pete had left an opening through which you could drive a Mack truck, and there was obviously a Mack truck handy. I swung into the driver's seat and put the pedal to the metal. "Who was it who let Sweetlips escape in the first place?"

I have to give credit to Pete. He realized that bad blood was very likely to be spilled, and when bad blood is spilled, it is neither easily nor quickly mopped up. He grinned weakly and said, "Maybe it was your crow that got Lyle to the finals, and Lyle just blew it."

I, too, backed off my high horse. "It's just a pity that instead of making fifty thousand smackers or so, we're out $750 for a crow of dubious deductibility. And a twenty-buck bribe. All we have to show for it is half a pound of leftover birdseed."

Suddenly, Pete got the look on his face that meant he was either hungry or he had an idea. When he didn't immediately head for the kitchen, I knew he had an idea. After about a minute or so, he looked at me and said, "Freddy, we're a couple of idiots. There might be a way for us to make some money out of this. It's been right under our noses, and we didn't see it. And all it will cost us is a couple of phone calls."

Well, maybe you saw it, and if so, my hat is off to you. But just remember, you weren't here chasing after substitute crows and losing sleep during the World Series of Poker. It's a lot easier to pick up on it from the comfort of your armchair.

Pete got on the phone and called the publishers of the newsletter that had served as a vehicle for the letters predicting the horse race winners. As he had hoped, they had no idea what was happening. The idea had been dreamed up by two guys, one in the circulation department who had access to the computer with the mailing lists, and the other in shipping, who could stuff the newsletters.

A couple of weeks later, we got a grateful letter from the publisher, who said that a scam like this could have destroyed the credibility of the newsletter. Accompanying the letter was a token of his gratitude, consisting of a $5,000 check. It wasn't $100,000, but by then we had realized that getting our hands on $100,000 had been a huge longshot to begin with.

We've still got the remainder of the birdseed, but we're not keeping it as a memento. If you drop by the house in Brentwood, just ring the doorbell and ask for it, and it's yours.

CHAPTER 9

THE WINNING STREAK

It had been a while since I'd heard from Lisa. I'd left a couple of voice mails and hadn't heard back. Consequently, I wasn't in the best of moods.

I don't know about you, but when things are not going well in one of the departments of life, such as relationships, I always look for positive developments in one of life's other departments to offset this. At the moment, however, I wasn't finding any such positive developments.

Psychologists talk about displacement; when you're irritated with someone and for one reason or another you can't take out your irritation on that person, you look for someone else on whom to take out your irritation. So maybe when you hear what I have to say, you won't feel that I had any right to be sore at Pete. However, he was the nearest available subject for displaced irritation, and I felt that I had a valid reason to take out my irritation on him. I had been hoping to find positive developments in the business department to compensate for not hearing from Lisa—and for reasons that will soon be clear, that wasn't happening. So, as I wrote out my monthly rent check, I was more than just in a mood to displace irritation; I was coming close to all-out grudge mode.

No, he hadn't raised the rent. Nonetheless, it did have to do with cash flow. There is a natural tendency of cash to flow from the renter (me) to the landlord (Pete). This creates what is known as a negative cash flow to the renter. "Negative cash flow" is investment jargon for losing money, and I have always had bad feelings about losing money.

Currently, my primary source of positive cash flow was the work I do with Pete. At the moment, he was turning down cases left and right, leaving me twisting (financially) in the wind. You'd have been in grudge mode, too.

I don't know what it is about the football season that turns even seasoned sports bettors such as Pete into raving maniacs. During the baseball season, assuming that you can find a bookie or get to Las Vegas, you can place bets every day. It used to be easier to find a bookie by simply going to the Internet, but the government has clamped down on that, as well as on eliminating online poker. But betting on football is practically a national institution. The football season begins before the baseball season ends, so you might think that football simply presents a few additional opportunities to place bets on weekends. Well, you might think that, but the hardened sports junkie certainly doesn't.

The coming of the football season had turned Pete, as hardened a sports junkie as ever came down the pike, into a virtual vegetable. Except for occasional forays to (a) the refrigerator, (b) the bathroom, and (c) the bedroom, he had grown roots in front of the TV set. But I was used to this type of behavior, and besides, part of my job was to pry Pete away from the tube. However, right now it would have taken dynamite to accomplish this goal because Pete was on the mother of all winning streaks.

Hence my soreness. My cash was flowing negatively, while his was flowing positively. Flowing doesn't begin to describe it— it was positively gushing. And so was Pete. It was hard to remain civil while I was writing out rent checks, turning down cases, and hearing Pete gleefully announce that he had gone nine and two in the pros last Sunday and had won all his two- and three-unit bets.[1]

So when the latest phone call came in on the business line, it was the answer to my prayer. Before the call, I had been trying to dream up schemes to talk Pete into taking a case, and I had come up completely dry. Even if I had been sufficiently imaginative to invent the phone call I had just received, as you will soon see, I could never have acted upon it.

I walked into the living room, where Pete was sprawled on the couch in front of the big-screen HDTV. He looked up as I entered the room. "You look like the cat that just swallowed the canary, Freddy. What's up?"

"Maybe your number," I replied, pushing away the bag of tortilla chips he was shaking under my nose. "No, thanks. How do you stand with your bookie?"

Pete consulted a few slips of paper in front of him. "I'm current as of last Tuesday, but I'm burying him this weekend. Why?"

"Your bookie works for Victor March, doesn't he?"

"Yeah, March gobbled up a lot of the independents on the Westside. It's hard to find good lines anymore." He crunched a few more tortilla chips and washed them down with some Coors. "I say again—why do you ask?"

"Because Victor March wants to see us this evening. I know your reluctance to take cases during the football season, but I accepted anyway. If I didn't, maybe he'd cut off your credit. Then whom would you bet with?"

Pete gazed at the living room. "Maybe we'd better straighten up the place before he arrives. This place is a mess."

I snorted. "Victor March doesn't arrive. He sends out. A chauffeur will pick us up at eight this evening."

Pete was still not interested in taking cases. "I wanted to watch the game this evening," he stated, aggravating me even further. "I'm loaded up on UCLA minus six over Arizona State."

"Why don't you record it on the DVR? Isn't that why DVRs were invented?"

"It's not the same thing as seeing it live," Pete said assertively. "Besides, there's a universal conspiracy on the part of the world to inform you of the outcome of any sports event you have decided to record. Particularly if you've got a heavy bet down."

"Think of it this way. If we get back early, you can fast-forward through the game and avoid those annoying commercials. If we get back late, instead of having to watch dull pregame commentary Sunday morning, you can fast-forward through the game during breakfast. Plus, there's always the chance that March may have something interesting and profitable for us. We really shouldn't pass it up."

Maybe my powers of persuasion were improving, or Pete just wanted to meet his bookie in person, as a little after eight we were in the office of High Rollers, Victor March's elegant Beverly Hills club. March himself was a little bit of a letdown, as he looked more like a florist than a nightclub proprietor and reputed head of a major bookmaking operation. He offered drinks, which Pete and I accepted. After a few preliminary remarks, obviously designed to create a cordial atmosphere, he came quickly to the point.

"The two of you have a reputation for being able to solve unusual problems," he said. "I have one that I think might be of interest to you." We looked interested (at least, I did—Pete has one of those faces that it's really hard to look interested with), and so March continued, "I'm out ten thousand bucks. I'm sure I've been swindled, but I have no idea how it was done. It's chump change, but if the word gets around town that I've been taken, I'm not going to look good. It's bad for me and bad for business. And if I find out how it's done, it's going to be bad for DiStefano as well." Suddenly March looked a lot meaner than my neighborhood florist—or any neighborhood's florist, for that matter. The atmosphere in the room had suddenly gotten a lot less cordial.

When business is being discussed, it is understood that my job is to stay alert. Pete's is to conduct the conversation. I stayed alert. Pete may have been suffering withdrawal symptoms, not having seen a pigskin in more than an hour and a half, but he kept his share of the bargain and asked March, "Do you mean Danny DiStefano?"

I was still alert but feeling a lot less enthusiastic about this situation than I had when I first received March's phone call. Danny DiStefano and Victor March probably accounted for three out of every four sports dollars bet on the Westside. However, where March used his profits to run legitimate enterprises (or so I'd been informed), Danny DiStefano was reputed to sink his back into drugs. DiStefano also was reputed to have more aggressive methods of debt collection than Victor March, as well as a squad of aggressive debt collectors whom you didn't want to annoy. I had to stop my teeth from chattering as I contemplated being caught in a Danny DiStefano–Victor March crossfire.

Pete seemed oblivious to such unfortunate consequences. It occurred to me that, despite his recent winning streak, the reference to $10,000 as chump change had impressed him. He finished the last of his drink and asked March to describe how he was swindled.

"I ran into DiStefano the other night in a club over on Melrose. We had a few drinks and then started making bets on whether the next person to enter the club would be a man or a woman." March paused, but that didn't surprise either me or Pete. Hardened gamblers have been known to bet on which raindrop would slide down a window and reach the bottom first.

After a short pause, filled only by silence, March continued. "After a few bets, he came up with an amazing proposition. DiStefano told me he was willing to lay me odds of eleven to ten that he could guess the sex of a person he had never met!"

Pete raised his eyebrows. "I'd almost be willing to go for that one myself. Are you sure that he had never met the individuals in question? That's the obvious way to rig the bet."

March took a puff of his cigarette. "That's what has me stumped. I would swear that he would have had no chance to do so. You see, I selected the people myself!"

It is hard to surprise Pete, but I could see that March had managed it. "Then you knew the sex of these people and still lost?" Pete asked.

"Don't be ridiculous," March snapped. "It was like this. He suggested that I just choose people at random, and he would guess the sex of their siblings. In order to insure that there would be no room for argument, we would only use those individuals who had precisely one brother or sister. It certainly seemed to me that the brother or sister was equally likely to be a man or a woman, and I was therefore receiving eleven-to-ten odds on an even-money proposition. Needless to say, I went for it, and it cost me ten thousand bucks. I dislike being swindled, especially by scum like DiStefano. It's worth five thousand to me to find out how he did it."

Pete got up. "A very interesting problem, Mr. March. Let us think about it overnight, and Mr. Carmichael will call you in the morning with our answer."

When we got back to the house, Pete immediately turned on the previously recorded UCLA–Arizona State game. I blew up.

"There's five thousand bucks on the line, Pete. To say nothing of the continued good will of our client, the influential and potentially dangerous Mr. March. Maybe you should get to work."

Pete watched the opening kickoff, then favored me with a reaction during the commercial. "Don't sweat it, Freddy. It's in the bag."

I had no idea why he was so optimistic. "You haven't asked anybody any questions. You haven't visited the scene of the crime. How can you tell me with a straight face that we've just made $5,000?"

As I mentioned at the start of this story, there are times when Pete drives me crazy. I have to admit, though, that when he decides to explain something, he does it quite clearly.

"Look, Freddy, here's what happened. Since the person being asked what the sex of his or her sibling was had only one sibling, there are four possible orders in which the two children were born. These are boy–boy, boy–girl, girl–boy, and girl–girl. Each of these is equally likely."

I thought about it for a minute. "Yeah, Pete, that makes sense. But I don't see what difference it makes."

Pete paused briefly. "Let's assume that March asked a man the sex of his sibling. Since he asked a man, the two children could not have been girl–girl. This leaves only three possible cases remaining: boy–boy, boy–girl, and girl–boy. In two out of three cases, the man's sibling was a girl. If my analysis is correct, DiStefano guessed that the sex of the sibling was opposite to the sex of the individual being asked. All we have to do is confirm that with March when we call him tomorrow."

(Probabilities for two-child family continued on p. 193)
I thought about it. "Pretty neat. So instead of laying 11 to 10 on an even-money proposition, DiStefano was laying 11 to 10 on a 2-to-1 favorite."

Pete nodded. "No wonder March got clobbered. If he bet $10 three times, he would receive eleven once and pay ten twice, assuming that the siblings ran true to form. This would represent a net loss of $9 on a $30 investment, or 30%. Even Vegas gives you better odds."

I shook my head in amazement. "I sure hope that March keeps his word. We haven't done much work for five thousand bucks."

Pete chuckled. "I'm sure March will keep his word. Especially when we tell him how we're going to make his money back. With interest."

"We're what?"

"You heard me, Freddy. Get March on the phone first thing to-morrow morning. Well, second thing. Wait 'til I'm up."

I didn't get much sleep that night, as I couldn't imagine for the life of me what Pete had in mind. As soon as I got hold of March, Pete got on the extension and took over the conversation. He first asked March if DiStefano had always bet that the sex of the sibling was opposite to the sex of the person March asked.

March thought a moment. "Yes, I'm sure it was. Why do you want to know?"

Pete explained his theory. March cursed.

"So that's how he did it. You've earned your money, Mr. Lennox, but I'm going to be the laughingstock of the city when they find out."

Pete's timing in these situations is pretty good. "What would it be worth to you to get your money back, Mr. March?"

March didn't answer directly. "You think you can do it? You can name your price if you do."

"Half of what you win back. And we'll cover your losses if you lose."

I almost dropped the phone. "Are you crazy?" I quickly mouthed to Pete.

When Pete gets one of his bright ideas, he can't be stopped by anything short of physical force. The only thing that prevented me from using physical force was the knowledge that his bright ideas had an unbroken record of creating positive cash flow. Plus the fact that Pete was bigger than I was, and other than hitting him over the head with a heavy object, I had no idea how to go about using physical force.

Anyway, Pete was now under full sail. "Yes, Mr. March, we'll guarantee your losses, up to $5,000." Well, at least I had had some effect. "That is, if you follow my instructions exactly."

March knew a good deal when he heard it. "Just tell me what they are, Mr. Lennox."

"The Beverly–Chatham Hotel is near your club, isn't it?"

March sounded puzzled. "Yes, it's just down the block."

"Then here's what you do. Invite DiStefano to dine at High Rollers this evening, and suggest that you feel the need to get your money back with the same bet. There's a good chance he'll go for it."

March snorted. "I'm sure he will. But why won't I continue to lose just as before?"

"Here's what you have to do. After the two of you have dined, maneuver him into the bar at the Beverly–Chatham Hotel on some pretext and continue to make the same bets you did last evening. Only this time, pick other people in the bar as the subjects of your bets. No matter what you do, stay there. To make sure that you adhere to these conditions, Mr. Carmichael will accompany you."

When working with an eccentric like Pete, you have to give him some space. Pete liked to give the impression of pulling rabbits out of hats, explaining the rationale behind his ideas after the plan worked. Ever the eccentric, Pete sent me to accompany March and DiStefano, deciding he wanted to watch the Sunday night NFL game on TV, thus completing an eye-straining triple-header of football. It was a wonder he still had eyeballs. I asked him if he had any instructions.

"Just keep track of our winnings, and make sure we don't get stiffed." I wish I had his confidence.

Dinner was an extremely interesting experience. The food and drink at High Rollers was superb. It was also an edifying experience, which I can appreciate a lot more in retrospect, watching the byplay between March and DiStefano, each of whom had a hidden agenda of hornswoggling the other.

We went out for a short walk afterward, suggested by yours truly, to work off a few of the calories, and then fate intervened on our behalf. The fine autumn day turned into a rather nasty evening, just as we were passing the Beverly–Chatham Hotel. It seemed obvious to duck in out of the rain, and there was certainly no point in going elsewhere, as the Beverly–Chatham Hotel had a fine bar. Very soon some heavy betting was under way.

Pete slept late next morning. No news there, as he sleeps an average of nine or ten hours a day, virtually guaranteeing that he will sleep late. He finally woke up, though, sauntered into the living room, and asked for the final tally.

With great pleasure, I slid over March's check for $5,000, plus $13,000 in cash! Pete nodded, as if it was no more nor less than he expected.

He might have expected it, but I didn't have a clue. I needed an explanation before I went crazy trying to dope it out.

"You should have been there, Pete. We must have won about five of every six bets we made last night. DiStefano almost had a stroke! How in the world did you do it?"

Pete grinned. "Pretty simple, actually. I really didn't do anything at all. But I did happen to read in the paper a few days ago that the International Society of Identical Twins was having its annual meeting this week, and that they had engaged most of the rooms in the Beverly–Chatham Hotel. As you may know, identical twins must have identical DNA sequences. Since sex determination is a part of that sequence, they must consequently be of the same sex."

It's hard to stay sore at a guy who has just put nine thousand bucks in your pocket for a couple of days' work. "Forgive and forget" is my motto, even though I wasn't going to tell Pete that all was forgiven because he's so oblivious that he might not even have known I was sore in the first place.

There were other positive developments on the cash flow front, at least as far as I was concerned. As all winning streaks must, Pete's came crashing back to Earth, just as he had increased the size of his betting unit. He was even still ahead on balance after a Monday night disaster featuring an eighty-three-yard run with a recovered fumble with forty seconds to go that snatched defeat from the jaws of victory. Events such as these reminded Pete of both his own fallibility and the desirability of positive cash flow. He even asked me if we had any potential business!

Needless to say, I was heartened. Then I thought of an almost surefire way to boost Pete's sagging spirits. I made a phone call the next morning and waited until Pete was fully awake to make my move.

"How'd you like to double this evening, Pete?" I asked. "Come on. It might take your mind off other matters."

"Mmm. Who's your date?"

"Arlene Halliburton."

"Don't know her. New girlfriend?"

"She's currently classified as a work in progress. Anyway, she's cute, and she's interested in fixing you up with her sister. I described you, and Arlene said that you sounded like the type of guy who appeals to her sister."

Pete pursed his lips. "Girls like to fix up their friends who can't get a date, Freddy. How do you know I will like her? Have you ever met her sister?"

"No I haven't, but I don't think you'll be disappointed. Both heredity and environment are working in your favor."

Pete looked puzzled. "Heredity and environment?"

"I met Arlene in the bar of the Beverly–Chatham Hotel."

Note: A surprise awaits the reader following the magnifying glass on p. 193!

CHAPTER 10

ONE LONG SEASON

I couldn't believe my ears. Pete had been really depressed for a week as a result of a large number of losing football and basketball bets. Well, that's the nature of risk-taking—winning streaks are often followed by losing streaks. It happens to Wall Street investors as well.

"Did I hear you right, Pete?"

"You heard me, Freddy. There is a meeting next Tuesday evening in Santa Monica of Gamblers Anonymous. I'm going to check it out. When I think of all the time I've spent studying racing forms and team histories, if I had put all that effort into something constructive I might have had something to show for it." He lumbered back to the main house, approximately three parts dejection and one part resolution.

I must admit I was nervous. When a person makes one major lifestyle change, maybe others are in the offing. What if Pete decided he wanted to accept a job somewhere other than in Los Angeles? I might find myself minus both a residence and a business partner. I consoled myself with the knowledge that Pete and inertia were more than just on speaking terms; they were really good buddies.

The status remained quo for a couple of days, and then the phone rang. After taking the message, I went to look for Pete. I found him in the living room, lying on the couch and watching a reality show. A reality show! Pete never watches reality shows. Desperate measures were called for.

"We've got a client, Pete. My accountant Angela recommended her. Rise and shine."

Well, at least I had prodded Pete out of the prone position. He sat up and said. "When?"

"About an hour or so. If you'll shave and put on some respectable clothes, I'll clean up the living room." I still have New York ideas about meetings with potential clients.

Thirty minutes later, both Pete and the house were in better shape. And a good thing it was, too, because as Julie Rydecki, the potential client, crossed the sill I could see Pete perk up as he realized that maybe there were other reasons for living besides football and basketball games. We offered a cup of coffee, she accepted, and soon we were discussing her situation.

Pete was as close to beaming as I had seen him in weeks. "And just what can we do for you, Julie?" I didn't think that he had forgotten that a client should initially be addressed as Ms. Rydecki; my guess was that he wanted to move to a first-name basis as soon as possible. Julie didn't seem to mind.

I knew what was coming, having had a brief outline from Angela, but Pete didn't, as it had taken him so long to shave, shower, and change that there had been no time for a briefing. So he was a little confused when Julie asked, "Have you ever seen *The Proud and the Passionate?*"

Pete shook his head. "I don't watch soap operas much."

"I'm an addict," Julie said. "I haven't missed an episode in the four years it's been on. In fact, that's why I'm here."

Pete's confusion deepened. However, one thing he had learned from me was never to show confusion to a client. "Maybe you'd better start at the beginning, Julie."

Julie rearranged herself more comfortably in the leather chair. "Almost a year ago, Silktex shampoo, which sponsors the show, started a promotion involving writing an essay about the show. First prize was $20,000 and a bit part in an upcoming episode. To make a long story short, I won!"

"Well, congratulations!" we both said in one breath. Twenty thousand dollars is not chopped liver, and it looked as if some of it might be headed our way.

Julie took another sip of her coffee. "Thank you. It was very exciting, being on the show. The episodes are taped months in advance of their actual showing, although occasionally they need to reshoot a scene, and they made me sign a nondisclosure agreement not to reveal anything about the show until after the season finale. That's the episode I'm in, and it runs next week.

"Of course, my part was very small, but I'm dying to see it. I'm also dying to see it for another reason."

"What's that?" Pete and I asked, again almost in one breath.

"One of the Madison Avenue types convinced Silktex a couple of weeks ago to try a stunt. They asked me to pick whom Debbie St. Clair was going to marry. If I was right, I would win $100,000."

"Who's Debbie St. Clair?" We were no longer in sync, as only Pete spoke this time. Having watched *P&P*, as it is called in the blogs, I at least knew who Debbie St. Clair was.

"Debbie St. Clair is the heroine of the show. For more than a year, she has been pursued by three men. It's common knowledge that the season finale will involve her choosing whom she is going to marry. A few of the actors know because the episode was filmed months ago, but they're sworn to secrecy as well.

"There's Judson Wyatt, rich and powerful owner of a chain of radio and TV stations. Suitor number two is Bennett Ellison, a darkly romantic figure who also has some sort of hold over Debbie's father. It hasn't been completely spelled out, but it seems that Debbie's father has a background he is trying to conceal. Before coming to Madison, California, where the show is situated, he apparently changed his name, but Ellison found out about it."

Julie paused for breath and continued. "The third suitor is Ralph Lowell. He's an instructor at the university where Debbie is just graduating. He's head over heels in love with Debbie, and I think Debbie's actually in love with him. But if Debbie decides to marry Ralph, Judson Wyatt, who is on the Board of Regents, will try to get Lowell fired. And Ellison will threaten to tell everything that he knows about Debbie's father."

Pete shook his head. "I'd love to be able to help you, Julie, but it's not really my line of work. As far as I can tell, there's no reason for her to prefer one to another. They're all equally likely. It seems like it's just a guess."

Julie held out her cup, and I refilled it. After another sip, she resumed. "That's how I saw it, too, so I just went with gut instinct. I thought she'd marry Judson Wyatt. It wasn't a complete hunch. Two of the most successful soaps, *Dallas* and *Dynasty*, feature marriages to rich and powerful people. So I thought I'd go with Wyatt. After all, soap operas generally try to do what made previous soap operas successful."

"Sounds pretty sensible," Pete said. "But where do we come in?"

Julie drained the last of the coffee. "I'm coming to that. The man Debbie will marry will be revealed on next Tuesday's show. But I do know this. She's not going to marry Bennett Ellison."

"More gut instinct?"

"No, a major pile-up on Interstate 5. Just last episode, Bennett Ellison was involved in the crash, and he's up in some hospital in Monterey in a coma. He may die, or he may not regain consciousness. At any rate, Debbie told us last week that she hadn't planned on marrying Ellison, anyway."

"So now it's a choice between Wyatt and Lowell, and you're still in the running."

Julie nodded. "And here's where the sponsors' gimmick comes in. They have offered me the following deal. During the commercial break that comes with about fifteen minutes to go, they are going to put in a live phone call to me. For $5,000, I can switch my choice and say that Debbie will marry Lowell. Or I can pay no money and stand pat with Wyatt. The last fifteen minutes of the show will make clear whether I am a winner or a loser."

Julie settled back in her chair and continued. "As you can imagine, everyone I know has offered advice. Some tell me that it's fifty–fifty, and that since I could be right either way, why not save the $5,000 and stay with Wyatt? Others come up with a variety of reasons why I should switch. Anyway, Angela seems to think you might be able to come up with a good reason why I should do one or the other."

Pete closed his eyes and thought for a moment. When he opened them, he said, "Julie, I can tell you what you should do and why you should do it. But it's a little hard to decide what to charge you. Let me make you an offer. Our fee will be $5,000, but it will be contingent on your taking our advice and winning the $100,000. If you don't take our advice, or you don't win the $100,000, you pay us nothing. Does that sound fair?"

Julie looked at him. "Five thousand dollars is a lot of money."

Pete nodded. "But it's only a small chunk of $100,000, and it's probably deductible. Ask Angela."

Julie thought for a moment, and then said, "All right. What should I do?"

Pete was about to open his big mouth, but I got there just in time with a modified standard contract. Julie signed, and I nodded to Pete. "Go ahead." I must admit I was pretty interested in what he would say, and his reason for it.

"You should pay the $5,000 and switch to Lowell," Pete stated decisively.

Julie's eyes narrowed. "You'd better give me a real good reason for spending five thousand extra bucks."

Pete paused to organize his thoughts and then began. "First of all, suppose that instead of Debbie having three suitors, just imagine for the moment that she had a thousand, and that it was up to you to pick the one out of a thousand that she was going to marry. Wouldn't you agree that you'd have to be awfully lucky to pick the right one?"

Julie thought about it for a moment. "Yes, I would. But they didn't ask me to pick from a thousand possibles but from three."

"I know," Pete assented. "But it makes it easier to see the reasoning if there had been a thousand potential husbands. Anyway, suppose that after you made your choice, the producers wrote a scene in which 998 of the remaining 999 suitors all ended up in comas. Would you switch now?"

Julie's nose wrinkled from the effort of concentration. Then, all of a sudden, she lit up. I could almost see the cartoon lightbulb signifying "idea" flash above her head. "I see what you're saying. I would have had to have been tremendously lucky to have picked the right one originally. Nothing they do could change that."

"That's it exactly!" Pete enthused. "Your chances of being right originally were one in a thousand, and that hasn't changed. If Debbie's husband-to-be were among the other 999, obviously the producers would select the 998 non-husbands-to-be of those 999 men to be involved in the car crash. So by switching, your chances of winning go from 1 in a thousand to 999 in a thousand."

/ (Conditional probability continued on p. 198)

Julie looked like she was taking a final exam. I couldn't blame her, as it was certainly one of the more esoteric arguments I had ever seen Pete come up with. "So what you're telling me is that, with three possible suitors, my chances of being right originally were one in three, and if I switch my choice they'll go to two in three."

"Right again! In the long run, if you have two chances in three of winning $100,000 instead of one chance in three, it's worth more than thirty-three thousand. Admittedly, there's no long run here, but if I were in your shoes, I'd pay the $5,000 and switch in a flash. Especially since you're playing with house money. Even if you switch and it's wrong, you won't have to pay our fee, and you're $15,000 and an appearance on prime-time TV to the good."

Julie got out of her chair. "I've still got a few days until Tuesday. I'll think about it." She shook hands with both of us and departed.

"Well, what do you think?" I asked Pete.

"I think she understood what I was saying. She's a very bright girl. If she's rational, she'll switch." The enthusiasm of the moment was starting to fade from Pete's voice. "Considering the luck I've been having recently, though, she'll probably stick with her choice, figuring that she'll save $5,000."

"Well, we'll know Tuesday night. I'll cross my fingers."

By now it should be evident that a lot of interesting things were about to go down Tuesday night. As far as I knew, Pete was still planning to attend the meeting of Gamblers Anonymous Tuesday evening.

I got back to the house Tuesday evening at about 5:30. As I was taking off my jacket, the phone rang. I picked it up.

"Freddy? It's Pete. I want you to call my bookie and ask for the line on the UCLA–Washington basketball game. If UCLA is favored by four points or less, bet five hundred on the Bruins to win."

I was stunned. "I thought you were going to a meeting of Gamblers Anonymous, Pete."

"I am. But just in case it doesn't take, I've had this flash of incredible insight into the game. Sorry, Freddy. I've got to go. I'll be back in a couple of hours." I replaced the phone.

Mine not to reason why, etc. I made the phone call, found that I could get UCLA minus three and a half,[1] and placed the bet.

(See An Introduction to Sports Betting on pp. 231–33 and Notes to Chapter 10 on pp. 235–36.)

Two hours later, Pete arrived back at the house. It didn't look as if the meeting had "taken." The tip-off was at eight o'clock, and Pete settled down to watch.

At halftime the score was tied. Halftime was shortly before nine o'clock, and *The Proud and the Passionate* hit the airwaves at nine precisely. I reminded Pete of this fact.

"Don't worry, Freddy. I'll just flip back and forth between the two stations with the channel changer. Besides, if you recall, Julie said they'd call her up at about 9:45. We've got plenty of time."

I know what you're thinking. Why didn't we just record either the game or the show as a backup? Well, we tried, except that Pete had backlogged the DVR with so much stuff that we were out of storage space. I know that seems difficult to accomplish, but it was an older-model DVR. And, as Pete said, we could always change channels.

As you can well imagine, I was vastly more interested in the happenings on *P&P* than I was in a crummy basketball game. The next forty minutes or so passed incredibly slowly, at least for me. Pete, on the other hand, was glued intensely to the screen. Washington moved out by five early in the second half, but the Bruins put on a run and caught and passed them. With six seconds to go, UCLA was up by four, and Washington had the ball in their own backcourt.

Pete's teeth were clamped. "I'm doomed," he said through gritted teeth. "They'll just let them have an easy layup, and I'll lose the bet. Why does this keep happening to me?"

I thought of a thousand things to say and, in the interests of friendship and future business dealings, suppressed them. Washington brought the ball up court as time ticked off the clock. With two seconds to go, one of their players launched a rainbow jumper from beyond the three-point line. The buzzer sounded. The ball hit the backboard, rattled around the rim, and fell off. UCLA by four!

"Chalk up a W!" Pete shouted, jumping off the sofa, in a frenzy of excitement. The channel changer fell off his lap, hit the coffee table, and bounced onto the floor.

I looked at my watch. Quarter to ten. "It's 9:45! Change the channel!" I yelled at Pete.

He came back down to Earth. "Sure, Freddy." He picked up the remote and poked the button. Nothing happened.

On the screen, some commentator dressed in a sports jacket was interviewing a sweaty UCLA basketball player. I wanted to hear what Julie did and whom Debbie decided to marry.

"Gimme the remote!" I screamed at Pete. He shoved it at me. I poked at the "last channel" button. Nothing happened. I punched in the three-digit combination for the channel for *The Proud and the Passionate*. No response.

I'm no expert on electronics, but I pried off the back and looked at the batteries. They were leaking all over the place. "Have you got any AA batteries?" I yelled at Pete.

He checked. "No. We're all out."

I looked at my watch. 9:52. By now, everyone in the nation knew what Julie had decided to do. Everyone, that is, except us.

"Well, who do you know who watches *The Proud and the Passionate*?" I asked Pete.

He shook his head. "Just Julie. And I don't think it's advisable to call her up to get the dope. What about you, Freddy?"

I gritted my teeth. "We should have watched the show. You can get the sports scores any hour of the day or night, from the radio, the TV, the Internet, or one of your degenerate sports-crazy friends."

Frantically, Pete and I thumbed through our respective phone books, looking for likely soap opera watchers. All of a sudden, inspiration struck me. It may have struck you even sooner reading this story, but you didn't have five thousand bucks on the line.

"Of course!" I burst out. "Angela! She's Julie's friend, so she's certain to have watched the show."

Normally, I am somewhat reluctant to call people after ten p.m., but there are moments when exceptions have to be made. This was clearly one of those times. I placed the call.

All's well that ends well. Not only had Julie changed horses in midstream, but she had backed a winner. Debbie had decided to marry Ralph Lowell. Rumor had it that you could place bets in Vegas on how many episodes it would be before the marriage was in trouble.

I relayed the good news to Pete. He nodded contentedly and then reached for the phone. "Bernie? What's the line on the Knicks tomorrow night? Okay, gimme the Knicks for $200, and if it off² on the Lakers if I can get them at minus six or better." He replaced the receiver, contentment oozing from every pore.

"I guess it didn't take," I muttered sarcastically.

"What didn't take?"

"The Gamblers Anonymous meeting."

He sighed. "Well, Freddy, on the way back I got to thinking. It's nice to show a profit on the baseball season, and nice to show a profit on the football season, but one should always remember that life is one long season."

I couldn't argue with philosophy like that. Far-thinking businesses have the same view of their quarterly reports.

The other day Pete and I spent a chunk of the $5,000 on some new equipment. We now have a brand new big-screen TV in the living room. With a fancy channel changer, and picture-in-picture so we can watch two shows simultaneously, in case something like this ever happens again.

Oh, yes. We also got a smaller TV, ten bucks' worth of AA batteries, and a stand-alone DVR in addition to the one supplied by the cable company. Speaking of the cable company, we rolled over our cable plan for two years, which entitled us to a DVR with massive storage.

CHAPTER 11

THE GREAT
BASKETBALL FIX

A little of Ollie Richardson, used car salesman, goes a long way. Unfortunately, at six feet four and closing in on the weight of an offensive lineman in the NFL, there is a lot of Ollie Richardson.

Ollie is a regular at Pete's Tuesday night poker game, held on Tuesday night in order not to interfere with Monday night football. In addition to being as obnoxious as used car salesmen are generally reputed to be, Ollie is also a world-class gloater. This, of course, raises anyone's obnoxiousness quotient. He does, however, have some saving virtues. When the recent budget cuts at L.A. Unified School District threatened to scrap girls' basketball, Ollie stepped in and coached the girls' team at MacMillan Junior High, where his daughter was a point guard. MacMillan was winning, and Ollie was gloating.

You now have the same background that I did a month ago. It was then that Pete received a call from his Aunt Harriet, a soft-spoken lady with an unkempt halo of light gray hair, who was somewhat given to dithering. It seemed that Aunt Harriet's daughter Irene attended Rutherford Junior High, which was in the same district as MacMillan, coached by the one and only Ollie, with whom you are already acquainted. The budget axe had wreaked havoc upon the extracurricular activities at Rutherford as well, and the girls' basketball team was looking for someone to coach them. Would Pete be willing?

Are you kidding? Within even the casual sports fan lurks the firm conviction that every manager or coach is a complete dunce, and that if the aforementioned fan were merely placed in charge of the team, he or she could lead the lowly out of the cellar, the mediocre to championship contention, and the good to a dynasty. And Pete was more than just a casual sports fan. Pete knew that it was unlikely that management would put him in charge of the Dodgers or the Lakers but, as he said to me, you have to start somewhere.

It turned out that even though Pete had agreed to buy a pig in a poke, he had inherited a team that was closer to bacon than to hog jowls. The Rutherford Lady Basketeers had one major asset, a young lady by the name of Theresa Middlebury. At five feet seven, Theresa could be termed a giantess, at least for a thirteen-year-old. Even better, she could not only play good defense but also had a virtually unstoppable jumper, accurate from about twelve feet or closer. After the first practice session, Pete came back humming.

"How'd it go, coach?" I asked him.

Pete was clearly a happy camper. "They're reasonably talented, and they work hard. That, plus my knowledge of basketball, should prove to be a virtually unbeatable combination."

I thought it was time to inject a note of reality into the proceedings. "Pete, you've never coached a day of basketball in your life."

"Remember what I told you about my father, Freddy? He was a player. Besides, I've watched thousands of basketball games."

"Maybe millions. But it's easy to coach from the sidelines."

He digested this and nodded. "Nonetheless, coaching basketball is not rocket science." How could I argue with that?

I have to hand it to Pete. Either he was correct in his assessment of the talent and dedication of his young charges, or he really was a good basketball coach, or maybe a combination of both. It soon became evident that Rutherford and MacMillan were the class of the league and were on a collision course for the league championship. This rapidly became Topic A, or at least Topic B or C, at the Tuesday night poker game. The animosity that existed between Ollie and Pete (and between Ollie and everyone else, for that matter) induced spirited betting among the poker players. Both teams were undefeated, and since the game was to be held on Rutherford's home court, they had been installed as a four-point favorite. It was felt that over–under betting[1] would be out of place in a girls' junior high basketball game.

/ (See Notes to Chapter 11 on pp. 236–37.)

As the day of the game approached, I could see the level of intensity pick up. Pete had invited me to practices, doubtless to have someone there as a towel boy, but even I was getting involved. It was clear that the girls were going to give it their all. Although MacMillan did not have a Theresa Middlebury in the low post, scouting reports (faithfully relayed to us by Irene, who was privy to all the inside dope) said that they had two five-feet-four forwards who could really crash the boards. This was bad news, as Theresa was also Rutherford's best rebounder. On the other hand, MacMillan did not have a true point guard and was consequently a little weak on ball handling. Deciding that Irene's information was the real McCoy, Pete decided to teach the girls an aggressive trapping defense in order to exert maximum pressure when MacMillan was trying to get out of its own backcourt.

Pete had decided to use me as a sounding board because (a) I was developing a certain expertise and (b) I was usually within easy reach. After a recent practice session emphasizing Pete's new defense, he downloaded some basketball theory to me as we drove back to the house.

"You see, Freddy, in a game with only six-minute quarters, turnovers can be critical. I'm also planning on sending Theresa to the offensive basket whenever it seems that the trap is succeeding."

"Sounds good to me." I paused for a moment, deciding how to phrase my next remark. "Pete, I'm very impressed with the job you are doing. Do you suppose you could find someone at the poker game willing to take a hundred-dollar bet on the game? On Rutherford, of course."

Pete chuckled. "Let me give you Arnie Schrafft's number." Arnie was one of the Tuesday night regulars. "He does a little bookmaking on the side. Confidentially, he told me he's getting almost as much action on this game from the other Tuesday night poker players as he does on the USC–UCLA football game. Admittedly, USC has dominated in the past few years, which slows down the action."

I looked a little shocked. "Doesn't Arnie work for the school district?"

Pete nodded. "The budget cuts reduced his salary as well, so he's making a few extra bucks." He ruminated for a moment. "I think the line's four at the moment, though it's been jumping up and down a bit." We stopped for a red light. "I must say, Freddy, I'm happy to see that you approve of my coaching strategy sufficiently to back it with hard cash."

"Of course." I refrained from telling Pete that it wasn't simply the soundness of his strategy that had made me interested in risking a hundred bucks. At the latest practice, I had seen Theresa stick eight straight jumpers from the twelve- to fifteen-foot range. Not to mention twenty-two consecutive free throws.

By coincidence, I was asked to fill in at the next Tuesday night poker game. The atmosphere was certainly becoming charged. Pete had almost come to blows with Ollie, who had sneered at Pete's coaching ability, pointedly mentioning that his own team had achieved its record without benefit of a beanpole like Theresa Middlebury. Now, there is no question that Theresa Middlebury was taller than the average thirteen-year-old girl, but she was a little sensitive about her appearance, as are most thirteen-year-olds. Pete told Ollie that if he said anything like that to Theresa during the game, he would be liable to end up with a black eye, or worse. Ollie may have outweighed Pete by nearly seventy pounds, but Pete was about six feet two with long arms and was considerably more mobile than Ollie.

Practices were going extremely well, and it was beginning to look like the team was peaking at just the right moment. Then, three days before the game, disaster struck.

I was blissfully unaware that disaster had struck, as it was about ten in the morning. I had just finished breakfast and had taken coffee and the morning paper into the living room of the guesthouse. Maybe I was getting a little greedy, but I had decided to put another hundred on the game. I looked up Arnie Schrafft's number and gave him a call. He may not have been the only bookie in town, but even when Internet betting was legal, you couldn't find any lines for girls' junior high basketball games.

Following Pete's advice, I asked what the line was before I placed my bet. When I heard, the coffee I was sipping went down my windpipe the wrong way, and I nearly choked.

Pete has made it abundantly clear that, when he is sleeping, he is to be awakened only if it is extremely urgent. It struck me that this probably qualified. I buzzed him in his bedroom. It took eight rings before he answered.

"Yeah? What is it?" He knew it was me on the intercom, so he could afford to be rude.

"Pete, something has happened. I don't know what's going on, but I just called Arnie Schrafft to put another hundred on the game. He told me that the line was now MacMillan minus two. I thought I ought to tell you."

Pete whistled. "You're kidding! I hope you didn't fall into the trap of grabbing it. It's probably a sucker bet, although I have no idea why the line would change like that."

"No, I was stunned. I just thanked him and got off the phone."

Pete was silent for a moment. Then he said, "The only thing I can think of that might cause that violent a swing would be an injury of some sort. Arnie works for LAUSD, so he's privy to information we might not have. Well, we'll find out at practice this afternoon."

When practice arrived, we were relieved to find that there were no injuries—at least, none that we could detect. However, it seemed that something had happened to Theresa. She was throwing up bricks, her defense was off, and her mind was definitely not on basketball. Pete told me to take her to the free throw line and have her shoot a hundred free throws and keep track of how many she made. I did so and reported back to Pete.

"Sixty-six."

He consulted some numbers on a piece of paper and closed his eyes a moment. When he opened them, he said, "Freddy, you're an investigator, and we've got a situation that needs investigation. Something's happened to Theresa. Find out what's going on."

I was a little surprised. "Sinking sixty-six free throws out of a hundred isn't a crime."

"It is for Theresa. She's an 80% foul shooter, and sinking only sixty-six out of a hundred is three and a half standard deviations below the mean. Maybe one chance in a thousand."

He lost me. "Say what, Pete?"

"Freddy, if you were to keep track of how an 80% foul shooter does on a hundred free throws, it would form a bell-shaped curve. Theresa's success probability for a single free throw is 0.8, and her failure probability is 0.2. You would expect Theresa to sink 80%, which is eighty out of one hundred. That's the mean. To compute the standard deviation, you multiply 100 times 0.8 times 0.2, and take the square root of that number. So 100 is the number of free throws, 0.8 is the probability that Theresa will make a free throw, and the 0.2 is the probability that she'll miss. Let's see, 100 times 0.8 times 0.2 is 16, and the square root of 16 is 4."

I've learned to humor Pete's forays into mathematics. After all, I've profited big-time from some of them. "What good does that do you, Pete?"

"All bell-shaped curves have the same shape. There's a 68% probability of being within one standard deviation of the mean. Since the mean is eighty and one standard deviation is four, that means that 68% of the time that Theresa shoots a hundred free throws, she'll sink between seventy-six and eighty four. And 95% of the time, she'll be within two standard deviations of the mean and sink between seventy-two and eighty-eight. Only about once in five hundred times will she be further away from the mean than three standard deviations. This time she's fourteen free throws below her average, and 14 divided by 4 is 3½. So she's three and a half standard deviations below her mean. You don't get numbers like that by blind chance, Freddy. Something's going on."

✐ (Binomial distribution continued on p. 208)

He took the basketball I was holding. "I'll take care of the coaching, Freddy. Find out what's happening, or we're all in the soup."

My years of training had not been wasted. When you want to know something about someone, the best way is to ask a friend. I called Pete's cousin Irene.

It was a piece of cake. I herewith report our conversation in full.

"Irene, has something happened to Theresa?"

"Didn't she tell you? She won't be able to play in the game Saturday!"

"Did something happen to her, Irene? Has she had an accident?"

"No, she's fine. But the most amazing thing happened. She got a phone call last evening, and now she can go see the KrystalVision concert Saturday night!"

I read the entertainment pages, so I knew that KrystalVision was hot, especially if you were a thirteen-year-old girl. "I thought that concert was sold out, Irene."

"It was. But somebody said Theresa won two tickets in a rock station giveaway, and she told me I could have the other one."

I grasped at a straw. "But, Irene! Think of your teammates. You'll let them down terribly."

"No problemo. I told them all about it and promised them all autographed pictures of KrystalVision." She rang off.

I resolved then and there to limit both my investments on, and my future dealings with, teenage girls—at least with girls in the thirteen- to fifteen-year age group. Pete gritted his teeth when I reported.

"This is Ollie's doing, Freddy. I just know it. One of his relatives is a ticket scalper. Well, it's water under the bridge, I'm afraid. We'll just have to do the best we can."

There went a hundred bucks. Basketball practice the next day was dismal. Theresa and Irene showed up, but their heads and hearts were obviously somewhere else. Thursday's practice was every bit as depressing. Even if, by some miracle, KrystalVision canceled their performance on Saturday, Theresa would be so upset that she might cease to be an offensive force. She no longer seemed to be an offensive force, anyway, as she still couldn't throw a pea in the ocean.

I had mentally written off the hundred bucks and resigned myself to having to endure Ollie's gloating, to say nothing of the loss of a C-note, when a bolt from the blue arrived Friday morning. This particular bolt took the form of a bulletin from the District School Board, a copy of which, addressed to Pete, arrived in the morning mail. Pete was wolfing down the Friday brunch special at Irv's Deli, so I took the liberty of opening it.

The bulletin announced that disturbing events had come to the board's attention, and as a result it was forced to cancel the championship game on Saturday night. The board would have preferred to postpone the game, but the school term was coming to an end, and so there would be no opportunity to play the game at a later date. Rutherford and MacMillan were being declared cochampions, and the awards banquet would go on as scheduled. The last paragraph mentioned that a school district representative would go to each school in the district next week to give a lecture entitled "The Evils of Gambling."

Curiouser and curiouser. If nothing else, though, I felt that Pete would be somewhat relieved not to face the ignominy of losing the championship game to Ollie. Of even more importance, I could now be assured of not losing a hundred bucks. When Pete arrived back from Irv's Deli, I handed him the announcement.

He scanned it quickly. "I sort of expected something like this, Freddy."

Pete always gives me the impression that, on the day that the sun finally rises in the west, he will say, "I sort of expected something like this, Freddy." While I was certainly relieved to have the hundred bucks back in my pocket, my curiosity was itching.

"How on Earth could you possibly have predicted that the school district would cancel the game?"

"I couldn't be sure. However, you know that Arnie Schrafft was booking the game."

"So?"

"You remember that the line all of a sudden went from Mac-Millan plus four to MacMillan minus two? Well, that had to be due to a lot of money coming in on MacMillan. Arnie's fairly new to bookmaking, and he did the obvious thing, raising the price on MacMillan to attract more betting on Rutherford."

"That certainly makes sense."

"It makes sense, but there's a huge risk. Arnie could get seriously middled.[2] If the game ended with MacMillan either winning by one or losing by three or less, he'd have to pay off everyone who bet on MacMillan while the line on MacMillan was at plus four, as well as everyone who bet on Rutherford later on and got two points. And I'm pretty sure that a large chunk of his action came from Ollie betting on MacMillan plus four." He paused. "When that happens to a big-league bookie, they just lay off the extra action. Arnie's a small-timer, and where could you lay off action on a girl's junior high basketball game? I didn't think he could stand the pressure of having to pay almost everyone if the game ended unfavorably for him."

"So what do you think he did?"

"I'm virtually certain I know what he did. He let his superiors know that there was betting going on and that one of the players had taken a bribe—even though I'm not sure that Theresa accepting concert tickets constitutes a bribe. She's probably completely unaware of what's been going on behind the scenes." He looked thoughtful. "By the way, Freddy, do you suppose you could trace those tickets?"

"What tickets?"

"The ones Theresa got. If I could show they came from Ollie . . ." His voice trailed off.

I gave it a thought. "It depends. But is it really worth it? And even if I could, it's not a crime to send tickets to a rock concert to a minor." I thought about it for a moment. "Although maybe it would be a better world if it were."

"At least, a quieter one."

CHAPTER 12

IT'S ALL IN
THE GAME

"Is Señor Lennox at home?"

I couldn't help but feel that I recognized the slender, fortyish woman with the slight Latino accent on our doorstep, but I couldn't immediately place her. It was one of those situations where you normally see someone in one particular location, and then when you run into him or her elsewhere, you do a double take. This particular double take took about five seconds, and then I had it.

"Come on in, Dolores. I'll take a look."

Whenever you read about Los Angeles, you read about Hollywood, or sports franchises, or riots, or gang problems, but you never read about the people like Dolores who make up Los Angeles, and that's a pity. Twenty years ago, Dolores and her husband had just started Casa Dolores, a small Mexican restaurant on the Brentwood–Santa Monica border, when he was killed in a traffic accident, leaving Dolores to care for Maria and Pedro, their two children. In those twenty years, Dolores not only managed to make a success of the restaurant but also put aside enough to make sure that they each had a college education. Maria is doing her residency at the University of California, San Francisco, and Pedro is about halfway through USC. We're regulars at Casa Dolores, and Pete told me that when Pedro and Maria were in high school, they used to wait tables and clean up far into the night. Maria was a whiz at science courses, but Pedro had a little trouble with algebra, and Pete would tutor him in exchange for free meals.

I had gotten Dolores settled in the front room when I was spared the problem of locating Pete, who had evidently heard the door opening. It wasn't a double-take situation for him, possibly because Dolores had brought Pedro over evenings for tutoring. He greeted her warmly.

"It's nice to see you again, Dolores. I don't have to ask you how the restaurant's doing; it's always packed. How are Pedro and Maria?"

"They are both well, Señor. Maria, she will soon become a doctor!" Dolores pronounced the 'r' in Maria a lot more attractively than Pete.

"I'm not surprised. Tell me, Dolores, what can we do for you?"

She hesitated a moment. Then she took out a folded piece of paper and passed it to Pete. "I would like to consult you about what I should do about this."

The paper was a black-and-white copy of the front of a baseball card. It was a picture of Babe Ruth. On the front of the card were two dated autographs—one made by Babe Ruth in 1927 and the other by Roger Maris in 1961. Pete whistled.

"If this is yours, Dolores, put it in a safe deposit box quick! If the autographs are genuine, they are dated on the days that Ruth hit his sixtieth home run and Maris his sixty-first. A collector would give a fortune for it."

"They are genuine, Señor. And the card is safe." Dolores paused. "It was left to me by my grandfather. His father, my great-grandfather, used to play baseball in the Mexican leagues, and he played against Babe Ruth. They went out drinking together. My great-grandfather, he was watching the day Babe Ruth hit his home run, and he got Babe Ruth to sign the card. He was also watching when Señor Maris was playing, and he waited for three hours after the game to get him to sign the card."

"Then what do you need me for, Dolores?"

"I want to sell it. You see, I have investigated how to sell a card like this. One does it through auctions. There are two places in town where I could sell this card."

Pete nodded. "I'll bet I can guess. Classic Collectibles and Vintage Memories."

Dolores smiled. "I knew you would be helpful, Señor. Those are the two places that handle cards like this. I looked at the records of their auctions for the past ten years."

"You're very thorough, Dolores."

She was serious. "It is a business, Señor. Like running the restaurant. You must have good suppliers at a fair price. But with suppliers, you can give them a tryout. If they give you good food at a fair price, then you give them a longer contract. Here it is different. I only have one card to sell."

She took out another sheet of paper. "Each auction house has what they call a reserve price. It is a minimum bid. I have talked to the owners of both places. Classic Collectibles will set a reserve price of $20,000, and Vintage Memories is willing to set one for $30,000."

It didn't look like much of a problem to me. But then Dolores continued, and I saw where the problem was.

"In looking at the auction results, Señor, I noticed that the high-priced items sell for more at Classic Collectibles. They sometimes receive prices of $100,000 for their items, but the highest I could find at Vintage Memories was about $70,000." She paused. "I have been looking at these numbers so long I am going crazy. Some mornings I wake up feeling happy, and I know I am going to get lots of money for the card. So I feel I should auction it at Classic Collectibles. Some mornings I wake up unhappy, and I say to myself $30,000 is a lot of money. So I feel I should auction it at Vintage Memories. After going round and round for a week, I thought of you helping Pedro with algebra, and how smart Pedro says you are."

Pete nodded. "I think I can help you, Dolores." He took a piece of paper and made a drawing, which I have reproduced here:

	High Price (thousands)	Reserve Price (thousands)
Classic	100	20
Vintage	70	30

"I think the above diagram summarizes the information you gave me," Pete asserted. "The numbers represent how many thousand dollars you expect to get. For instance, if you auction the card at Vintage, and you get a high price, you told me that you might make as much as $70,000."

Dolores and I both studied the diagram. "That's right," she said. "But how does that tell me what to do?"

"You should auction the card at Vintage," Pete said, "for no matter what happens, you can assure yourself of $30,000, and if good things happen, you might make as much as $70,000."

"But I knew that," she protested. "Why shouldn't I sell it at Classic? I could make $100,000!"

Pete thought a minute. "Here's how to see it. Suppose that only one person wants to buy the card. If you auction it through Classic, it will only cost $20,000. If you auction it through Vintage, it will cost $30,000. If more than one person wants to buy it, then you will be happy to let them fight it out and make you more money."

Dolores was still looking dubious. "Here's another way to look at it," Pete said. "Suppose you had a lot of these cards to auction. Imagine that the buyers realize that there are enough cards to make them all happy. What would they do then?"

Dolores thought for a bit. "I see, Señor. They would make an agreement among themselves to buy all the cards at the lowest possible price. That's what I do when I have several competing suppliers. Now I see what to do and why I should do it."

She reached inside her purse and took out a checkbook. "You have given me good advice, Señor Lennox, and I will do as you suggest. What do I owe you?"

Pete looked at me. "Do you have any plans this evening, Freddy?"

I knew where he was heading. "None I couldn't cancel."

"Then cancel them." He turned to Dolores. "I'm just tutoring you instead of Pedro. You owe Freddy and me dinner tonight. And a bottle of sangria."

"You are most generous, Señor. You may have saved me $10,000. Dinner is a small price to pay."

Pete smiled. "We're going to be *very* hungry." We all shook hands, and she left.

When she had gone, I took another look at the diagram that Pete had drawn. "Is it always right for her to auction the card at the place that offers the highest reserve price?" I asked.

"In a situation like this, yes," he replied. "But there are other situations in which the picture isn't so clear."

A few weeks later, we were dining at Casa Dolores, and Dolores came over and told us she had sold the card for $62,000. She was so delighted that our dinners there are on the house until we start abusing the privilege, so I think we'll be eating a lot of Mexican food. While I was happy with the result, I was ready to kick my parents all the way back to Altoona, Pennsylvania. Our family had moved from there when I was about twelve, and when we moved, they made me leave three shoeboxes of baseball cards behind!

I've picked up a few pointers from Pete (and vice versa, I might add), and I've discovered that some of the things I learn from watching him have an odd way of showing up in other places. A couple of weeks later, I was confronted with a personal problem. It kept gnawing at me, and after thinking about it, I decided I might try to apply Pete's methods to it. But I couldn't get it to work, so I figured I must be doing something wrong. Pete and I must have been on some sort of psychic wavelength because when I opened the door, there he was. He was just about to open his mouth, but I got the first word in edgewise.

"Pete, how'd you like to help a buddy and business partner?"

He looked at me. "What kind of help?"

"It's probably right up your alley. Do you remember that analysis you did for Dolores a few days ago?"

"Of course. Speaking of Dolores, are you hungry?"

"Not just now. I've got a problem. I haven't heard from Lisa in a while, and so I was trying to decide whether to call her. It's like that old song: Once in a while she won't call, but it's all in the game. It occurred to me that I had two choices: to call or not to call. And there are two possibilities for her: either she wants me to call, or she doesn't."

Pete thought for a second. "Well, if you want to apply game theory, you have to make up a chart like the one I did for Dolores."

I nodded. "Yeah, I know. I've given it a little thought, and here it is." I shoved a piece of paper at him, which I've reproduced below.

"I guess I'd describe the numbers as how happy I feel I would be on a scale of one to ten."

Does Lisa want me to call?

	Yes	No
I call	10	0
I don't call	2	7

Pete looked at the diagram. "I think I see what you're saying. Obviously, the best case is that you call and she wants to hear from you. The worst case is that you call and she doesn't want to hear from you. Not only is it embarrassing, but it looks like your relationship is down the drain."

I nodded. "That's right. And if I don't call and she wants to hear from me, I've made the wrong move. Lisa feels that when a relationship is good, each of you knows how the other is feeling. Even though I'd have the wrong take in this situation, at least she wants to hear from me. That's some consolation. Finally, if she doesn't want to hear from me and I don't call, at least I avoid looking like an idiot, and I can always hope that her mood will change."

"Okay." Pete thought for a moment. "You've got a second hand on your watch, don't you, Freddy?"

"No, it's digital. Why?"

"That's okay. Look at it, and tell me how many seconds it reads."

I certainly didn't know what he had in mind, but I followed instructions. "Fourteen seconds."

"In that case, call her."

I looked at him. "Are you trying to be funny?"

Pete shook his head emphatically. "Not a bit. Game theory dictates that you should call her one time out of three but make your decision on a random basis. If your watch had read between zero seconds and twenty seconds, as it did, I would have advised that you call her. Otherwise, I would have suggested that you don't call."

I can't say that I was happy about my love life depending upon the second hand of a clock. "Perhaps you could explain a little further, Pete, without going into the gory details."

He sighed. "I'll have to go into some details. Suppose she wants to hear from you, and you have three opportunities to call. If you call once and don't call twice, you get ten points the time you call and two points each of the two times you don't call, for a total of fourteen points. Got that?"

"I can handle that 14 is 10 plus 2 times 2."

"Okay. Now if she doesn't want you to call, you get zero points the one time you call, and seven points each of the two times you don't. Again, a total of fourteen points. So, no matter how she's feeling, in the long run calling one time out of three gives you the same reward. That comes out to 14/3 points per situation, or four and two-thirds points every time this comes up."

It took me a moment or so with pen and paper to digest all this. "Very ingenious, Pete. But why don't I just call every time and hope she wants to hear from me?"

"You could do that. But suppose that she never wants to hear from you. In this case, you never get any points. However, by adopting the strategy of calling once every three times, and using a randomizing device to select whether you call or not, you have guaranteed a long-term average of four and two-thirds points per opportunity, no matter how she feels about your calling."

🖉 (2 × 2 games continued on p. 214)

I won't say that I followed it all perfectly. On the other hand, Dolores scored sixty-some thousand bucks following Pete's advice, and it cost her a bunch of Mexican dinners. I was getting the same advice for free. So, with some trepidation, I lifted the receiver and punched up the number.

It was with clammy palms that I heard her pick up the receiver. "'Lo."

"Hi, Lisa, it's Freddy."

"Freddy! You must have just gotten back. Pete said you were out."

"That's funny. He never said a thing about you're having called. Gee, it's good to talk to you. What's up?"

"Well, I may have some really good news about a new job opportunity, among other things. But it's still somewhat in limbo, and I don't want to jinx it by going into specifics. But cross your fingers and wish me luck."

"You know I always do, Lisa."

We went on from there to other topics of a more personal nature. Half an hour later, I signed off, feeling a lot more than four and two-thirds points to the good.

As soon as I hung up the phone, I had some urgent business to transact. I went downstairs, where I happened to catch Pete flipping the channel back and forth between pro and college basketball games.

"You know, Pete, I don't know whether to thank you or read you the riot act."

He paused in mid-channel-change. "Why would you want to read me the riot act?"

"When I spoke to Lisa, she said that she had called. You never gave me the message, and you knew she wanted to talk to me."

He got a little huffy. "In case you don't remember, Freddy, when you got back, I met you at the door. I was just opening my mouth to give you the message when *you* interrupted *me*. I never had a chance."

I thought back. He was right. Maybe I should back off my high horse. Then I realized that he had left my love life up to the digital readout of a watch.

"Okay," I said, "I'm willing to admit that I interrupted you, and that you would have told me Lisa wanted to hear from me. However, it has just occurred to me that I was very lucky that the digital readout on my watch was between zero and twenty seconds when you asked me to look at it. Had I read my watch some seven seconds later, you would have advised me not to call Lisa." I started to get mad all over again. "It seems to me you were willing to risk my love life just to get an opportunity to haul out one of your pet theories."

"Not a chance, Freddy. You were guaranteed to make that call."

I was a little surprised. "Do I remember correctly? I thought you said that, if the number of seconds read between zero and twenty seconds, I should call; otherwise, I shouldn't. And as I recall, the number of seconds read fourteen seconds."

"Your memory is fine. But I would have arranged things to make sure you called Lisa."

Silly me, thinking I understood game theory. "I thought the whole point of this game theory stuff was to randomize one's decision so as to get the same long-term result no matter whether Lisa wanted to hear from me or not."

"That's right. But I knew something you didn't, that Lisa had already called and wanted to hear from you."

I was still confused. "But what would have happened if I had looked at the readout seven seconds later? It would have read twenty-one seconds."

"Because then, I would have said that you should call if the readout had been between twenty and forty seconds. A priori and a posteriori probabilities differ, as I may have mentioned before. And seeing as you'd gone to all the trouble of preparing that chart, and you saw why the situation differed from the problem with Dolores, I didn't see how it would hurt you to learn a little more game theory." He paused. "Speaking of Dolores," he continued, "how would you feel about some Mexican food?"

Nothing stimulates the appetite like good news on the relationship front. The way to a man's stomach *is* through his heart.

CHAPTER 13

DIVISION OF LABOR

I hadn't known there was such a thing as the Bankers Club. Pete probably hadn't known there was such a thing as the Bankers Club. But our potential client, Ellis Packard, looked every inch a banker, and at about six feet four, there were a lot of inches. His suit was appropriately gray. His Mercedes was gray. And his retainer check for $5,000 was a particularly tasteful shade of gray.

I wasn't going to muff a job with a $5,000 retainer because of lack of hospitality, so I offered him a choice of refreshment. He settled for a Coors and launched into his presentation.

"I don't know whether either of you have ever heard of the Bankers Club," Packard began. We hadn't. "It's a club for members of the banking profession." We guessed that. "It has a lot of amenities, and it also gives its members a chance to network."

When I become dictator of the world, one of the things I intend to do is remove "network" from the verb list and place it back where it belongs, in with the nouns. Other items high on my agenda are to remove instant coffee and nondairy creamer from the world's grocery stores.

But I digress, partly because Packard had paused to take a drink, it being a hot and dry day. A few sips later, he was back in stride.

"I've been a member of the Bankers Club for a couple of years, and it's a good idea for bankers on the fast track to put in some time working on club committees. So I got appointed to the Elections Committee, which didn't seem like it was going to require an excessive amount of time or make me any enemies. Whenever you get on a committee, there's always a chance you can make some enemies. Have either of you ever served on a committee?"

Neither of us had had that dubious pleasure, and we said so. After finishing the remainder of his beer, Packard resumed his monologue.

"The election for the club president is held biannually, and in the past, if no member receives a majority, the leading two candidates have a runoff. Well, nobody likes runoff elections. There's more politicking and backbiting during this time. So I came up with what I thought was a very good idea. There were three candidates this year: Forrest Ackroyd, Helen Williams, and Artie Morris. I suggested that the ballots be printed up so that each person could indicate his or her first, second, and third choices. That way, it seemed to me we could avoid a runoff election."

At this stage, I thought I saw Pete wince, but evidently Packard didn't pick up on it. He squirmed around in his chair a bit. It occurred to me that he might not think it was seemly for a banker to ask for another beer, so I asked him if he wanted one. He nodded gratefully and continued after his vocal chords were sufficiently lubricated.

"Fifty-four ballots were cast, but there were only three different types of ballots that actually appeared." He handed us each a sheet of paper. "Here is a copy of the election results, which I posted on the club bulletin board after we had tallied them."

I'll present those results in a table, so you can look at them as the story progresses.

First Choice	Second Choice	Third Choice	Number of Ballots
Ackroyd	Morris	Williams	24
Williams	Morris	Ackroyd	18
Morris	Williams	Ackroyd	12

While we were scanning the results, Packard continued his tale. "It was only now that I was beginning to get an inkling that trouble might be looming on the horizon," he said. "There didn't seem to be a clear-cut winner. The Elections Committee wasn't scheduled to meet until the following week, so I figured we'd thrash it out then. I left the club, played a few games of racquetball, and returned later in the evening."

"I was sitting in the club bar with a Perrier, when suddenly I noticed that Forrest Ackroyd had occupied the seat next to me. Forrest is one of the more influential vice presidents of Transcontinental Trust. He patted me on the back, and to my surprise offered me an invitation to a golfing weekend at his home up near Monterey. We would play Pebble Beach, and I've always wanted to play Pebble Beach."

I could understand that. If you've ever seen a telecast of the golf tournament formerly known as the Crosby Clambake, you'd know why. I wasn't much of a golfer, but I wouldn't mind playing Pebble Beach myself. However, I seemed to recall that it's a closed club now, so you need an invitation.

Packard didn't waste any time. I made a note that if I ever needed a loan, I'd call him, for I didn't think I'd get any of this "the Loan Committee meets a week from Tuesday" stuff. "Forrest then observed that he hadn't actually seen a notice of who the winner was," he continued. "I told him that all we'd done was post the actual results of the balloting and that the Elections Committee was scheduled to meet next week. He then said that it seemed pretty clear that he had won the election, as he received the majority of the first-place votes, more than 40%. He told me I'd better get my short game in shape, as Pebble Beach was especially rough if you missed the green."

Pete was paying close attention, scanning the list of results while listening to Packard. Packard squirmed a bit more in his chair, though I wasn't sure whether it was because the day was muggy, or the chair was uncomfortable, or Packard was born to squirm. I guess people who are abnormally tall are continually forced to occupy chairs not designed for them.

He resumed the narrative. "The next day I was over at the club, and who should drop by during lunch but Helen Williams, who is the CEO at Consolidated Bankshares. She's very influential in local banking circles. The next thing I knew, I had been tendered an invitation to a party that is usually attended by the Who's Who in L.A. banking. I was astonished but gratefully accepted. Then guess what!"

At least one of us was up to speed, for Pete said, "I think I've got it doped out. Williams said that she expected to be declared the winner by the Elections Committee."

Packard assented. "That's what happened, all right. She said that, since Artie Morris was the first choice of less than a quarter of the voters, Artie couldn't possibly be elected president . Fortunately, since we had allowed each voter to indicate their preferences in order, and since thirty voters preferred her to Ackroyd, whereas only twenty-four preferred Ackroyd to her, it was clear that she should be declared the winner. She then said that she hoped I would enjoy the party."

More squirming and fidgeting. Since the day had cooled down considerably, and it was a $700 chair he was sitting in, I was inclining to the born-to-squirm theory, although the chair was so new that the leather was still a little stiff and uncomfortable.

Pete was still on top of things. "Let me guess what's next. I wouldn't be surprised if you were to hear Morris stake a claim."

"You've got a good feel for this, Mr. Lennox," Packard allowed. "I didn't actually hear from Morris, who works in the same bank that I do but is somewhat higher up. I heard from Celia Morris, his sister."

"Hold on a second," I interjected. "Is Celia Morris about five feet two, late twenties, long auburn hair, nice smile?"

"I take it you've met her," Packard declared.

"At a party a couple of weeks ago."

He looked a little embarrassed. "If you don't mind my asking, was she at the party alone?"

I reflected a moment. "I'm not sure. I seem to recall somebody with horn-rimmed glasses who seemed very interested in her."

Packard nodded grimly. "That sounds like Carroll Farnsworth. He's a CPA. What she sees in him I don't know." I guess, if you're the vice president of a bank, it's hard to see why a CPA would offer any competition.

"Maybe she has tax problems," I commented.

"Ahem!" Pete coughed. "Maybe we could get back to business."

"Actually, we were still on business," Packard asserted. "I've been thinking of asking Celia for a date, but I was under the impression she was still seeing Farnsworth. At least, that was what I thought until I received that phone call. Needless to say, I was delighted when Celia phoned. I was about to suggest dinner and a show when she mentioned how wonderful it was that Artie was to be the club president!"

"I tell you, I was floored. I asked her why she thought that, and she told me that Artie had told her that the club had a rather complicated ballot. However, when three points were awarded for a first-place vote, two points for a second, and one for a third, he had received 120 points. Helen got 102 points, and so did Forrest. Try as I might, Mr. Lennox, it's hard for me to envision Celia waiting up for Artie to bring her the results of the election so that she could go through that analysis."

Somebody had to say it, so I did. "You think her brother put her up to it."

Packard nodded. "She's in advertising and good at presenting data in the best possible light." He broke off, and then continued.

"You see my predicament. I want to go golfing at Pebble. It would really help my career to attend Helen Williams's party. And I would love to get something going with Celia Morris. But it seems to me that no matter how I handle it, I am not only going to lose two out of three opportunities, but I am going to make some enemies as well. I'm at my wit's end, Mr. Lennox. It's worth five thousand to me to get out of this with my weekend at Pebble, the Williams's party, and my chances with Celia Morris intact."

Despite the recession, bankers were still evidently doing quite nicely. However, it was time for me to butt in. Personal relations are my specialty, not Pete's.

I cast a longing glance at the retainer check, but you have to draw the line somewhere. "We're here to solve problems, but we're not a dating service."

Packard nodded. "Celia's my problem. It's just that she's tangled up with the election thing. Anyway, do you think you can figure something out?"

Pete stood up. "We'll let you know how it stands in a day or so, Mr. Packard. Thank you for coming." We shook hands, and Packard departed.

I felt optimistic. Packard's problem certainly dealt with numbers, and fortunately we had a house specialist in numbers.

"All right, Pete," I declared, "how does mathematics deal with this one? It seems to me that all three candidates can put up a pretty good argument."

Pete grimaced. "Yeah, they sure can. And even though mathematics has a lot to say about it, it can't do anything to solve the problem."

I was more than a little surprised. I was so used to Pete's pulling rabbits out of hats that I was expecting to see the tips of floppy little ears as one emerged.

"How can mathematics have something to say about a problem and not be able to solve it?" I demanded.

Pete shook his head. "There's a very important result known as Arrow's Impossibility Theorem. It won Kenneth Arrow the Nobel Prize for Economics in 1972."

It's really a shame that quiz shows are no longer in vogue because Pete knows these things the way kids know the lyrics of the latest hits. However, my attention was focused on the word "impossibility," which had an ominous ring to it. I commented on this.

"I don't like the sound of that, Pete. What is Arrow's Impossibility Theorem?"

"The exact statement of it is a little formal. But the upshot of the theorem is that it is impossible to devise a scheme for assigning preferences for a society such that the chosen scheme accurately incorporates the preferences of the individuals in that society. And that's precisely what we've got here."

(Voting methods continued on p. 219)

"I'm not sure that I fully understand, but it sounds like sayonara to five thousand or so bucks."

"Let me sleep on it." Pete's solution to practically everything is to sleep on it. He has told me that he believes that the purpose of the left side of the brain is to gather data on which to base a decision, and the purpose of the right side of the brain is to supply the insight to reach the correct decision. He has also told me he believes that sleep unlocks the hidden power of the right side of the brain. Me, I believe that he's just rationalizing the fact that he likes to sleep.

Well, I wasn't going to sleep on it. I like to think that problems are solved by working on them, not by sleeping on them. And it would be a feather in my cap if I solved one that Pete couldn't. So, while Pete was sleeping, I was thinking.

I had plenty of uninterrupted time for thinking, as Pete slept on this one for sixteen consecutive hours. When he had finally achieved sufficient consciousness, I could tell from the sour expression on his face that the much-vaunted right side had come up empty. He confessed as much.

"Sorry, Freddy. I guess we'd better call up Packard and explain the situation."

It was time to play my card. "That might not be necessary. I've been thinking about this, and I might have a way out."

Pete perked up, as anyone would when a $5,000 fee that one thought had flown out the window might be coming back home to roost. "Let's hear it."

"Ask Packard to give us copies of any material relating to the election, such as the club constitution and the actual ballots for starters. Maybe there's something somewhere that will invalidate the election."

Pete looked at me. "You may have something there. I wonder why I didn't think of that."

Praise from Caesar is praise indeed. "I think you probably got hung up by the thought that if a Nobel Prize winner couldn't solve it, how could you? Anyway, let's get Packard on the phone and see what happens."

A lawyer once told me that there was no such thing as an unbreakable contract, and I had spent a lot of time while I was working in New York poring over contracts. This one turned out to be a cinch. The ballots had been printed with no mention as to when the votes must be cast, which was specifically mandated by the club constitution. We called Packard, told him the good news, and recommended that he reissue the ballots in accordance with the club's prior practices.

A few days later, we received a very cheerful call from our erstwhile client. "Sorry this has to be such a short call," he bubbled, "but I'm in the middle of packing. I'm heading for Pebble Beach in another hour or so. But my invitation to the Williams's party stands. By the way, Freddy, you were right. It seems that Celia Morris broke up with her CPA, and I've got a dinner date!"

We were delighted to hear it. He was evidently so pleased with the outcome that he was going to send us another three thousand bucks. We were certainly not going to be so unmannerly as to refuse to accept it.

"You know, Pete," I reflected later, "I wonder what would have happened if we hadn't been able to find a loophole through which to wiggle."

"I can tell you that. We would have returned his retainer."

Returning retainers goes against everything I believe in. "Why would we have done that?"

"Freddy, when a mathematical theorem says that something's impossible, it is flat-out impossible. It's not impossible as in "the difficult we do today, the impossible takes a little longer." We would have been unable to solve his problem, at least from a mathematical standpoint. It's a good thing you figured out how to get around the impossible."

"Well, Pete, I'll tell you what. Maybe we can work out a new division of labor. You take care of the possible cases, and when an impossible one comes along, just let me know." I don't mind telling you, I was feeling more than a little smug.

Packard's check arrived a day later. Pete contemplated it with satisfaction.

"That's eight thousand added to the account, Freddy."

"Minus fifty for insurance," I said.

Pete looked blank. "Insurance?"

"Yeah, I thought a little insurance might be necessary. I didn't tell you the full story of the party where I met Celia Morris. She wasn't spending all her time at the party with that Farnsworth character. Maybe she really *did* have tax problems. At any rate, Farnsworth clearly seemed more interested in her than vice versa. Also, while I was getting my car, she got into a black Lexus and drove off alone. You don't have to be a detective to work out that she's almost certainly a free agent."

"She's listed in the phone directory," I continued, "so I took the liberty of sending her a dozen roses in Packard's name. In case you're unfamiliar with market conditions, that's fifty bucks' worth of roses."

CHAPTER 14

THE QUARTERBACK CONTROVERSY

'Twas the night before Christmas, but that didn't prevent creatures from stirring all through the house. Every other year, Pete gives a Christmas party, but this was the year to receive rather than to give, and we had received invitations to six different parties, scattered throughout the length and breadth of the entire L.A. area.

The bank balance was in good shape, and Pete had settled down to prepare for the forthcoming NFL playoffs and bowl games. So that he could concentrate on the pigskin finales, I had agreed to plan our party-going travels. I didn't give it a whole lot of thought because I thought traffic would be light on Christmas, so if I took us a little out of our way it wouldn't be a tragedy. It was late, and I was a little sleepy, but I made out a schedule and took it in to Pete for him to look at. He looked at it and didn't like what he saw.

"I think you could do better, Freddy."

As I said, I was sleepy. "It can't be so bad. I started with Harvey Davenport's party in Malibu and simply went to the nearest party after that. I just continued going to the nearest party until we were out of invitations."

He shook his head. "A lot of people will be on the road." I guess Pete and I had different views of Christmas traffic; but I had to admit he had more experience with what it might be like in L.A.

"I'd like to spend as little time as possible being stuck in traffic jams," he continued. "The 'nearest neighbor' algorithm can be very inefficient in dealing with traveling salesman problems."

I was still sleepy. "Say what?"

Pete put aside his football paperwork and took out a pen and a piece of paper. He then spent a moment writing down a table, which I have reproduced here:

	A	B	C	D
A	—	25	30	20
B	25	—	60	35
C	30	60	—	40
D	20	35	40	—

"Here's the idea. This table represents the distances between the four different locales. Anyway, suppose that we start from A. The nearest city to A is D, so we would travel 20 miles. From D, the nearest city not yet visited is B, which would be 35 miles. We would then be forced to go to C, a distance of 60 miles. That would be a total of 20 + 35 + 60 = 115 miles."

Now, nothing makes me sleepier than a deluge of numbers. However, I was pretty sleepy to begin with, so I thought if I paid attention, I'd get to go to sleep faster.

"I'm with you so far, Pete. And I think I see the trouble. We would like to avoid the 60 mile segment, wouldn't we? That's what creates the problems."

"That's it. It would be better to go from A to B, then from B to D, and finally from D to C. That would be 25 + 35 + 40 = 100 miles. In the absence of any other information, this would be a clearly superior route to take."

I yawned. "Well, I suppose you know a shortcut for figuring out the best route."

"Nope."

"What?" Hearing Pete confess that he didn't know how to solve a problem would have been almost as surprising as hearing him admit that he couldn't decide whom he liked in the upcoming football games.

"I don't know a shortcut. But no one does, not even the best mathematicians in the world. Other than simply listing all possible routes and calculating the total distance for each one, there is no known shortcut for selecting the optimal route. This is known as the Traveling Salesman Problem." He spoke the words as if they were in capital letters.

(Traveling Salesman Problem continued on p. 227)

"I'm shocked. But I'm also sleepy. If you don't like my planned itinerary, you might start by making up one of your own." I yawned again and turned to go. "Me, I'm going to hang my stocking by the chimney with care. Good night."

After opening our presents the next morning and having a late brunch, we embarked upon our all-day partython. As we prepared to depart, Pete handed me a computer printout.

"What's this?" I asked. "Another Christmas present?"

"I took your advice last night and wrote a computer program to list all possible routes and compute the mileage. This is the shortest route."

Since I had been elected to drive, I took a look. The game plan was to start at the bottom in Orange County, go north to Long Beach, go to a couple of parties in Beverly Hills, and close the day's festivities at Harvey Davenport's Malibu estate. Harvey was majority owner of the Raiders, who had recently moved—again—from Oakland to Los Angeles, where they were playing—again—at the Coliseum and where they still had—again—a devoted Los Angeles segment of Raider Nation. Since we had helped untangle a small personnel problem in one of his companies, we were on the guest list.

I was surprised that Davenport even remembered who we were. When we got there, it was a little surprising that Davenport even remembered who anybody was, since it was clear that no one had been monitoring his eggnog intake. He welcomed us enthusiastically. "If it isn't the sleuths! Merry Christmas!"

We wished him a Merry Christmas right back. At first I thought it was our greeting that triggered his reaction of peace on Earth, good will to detectives, but later events made it clear that more weight could be attached to the Raiders cap that Pete had the foresight to wear. He put an arm around Pete's shoulder and said, "What are you doing on Saturday, buddy?"

It didn't seem like a trick question, so Pete gave him a straightforward answer. "I'm going to camp in front of the tube and watch the wildcard games."

This remark did nothing to decrease Davenport's desire to spread peace on Earth and promote good will toward detectives. "Then this may be your lucky day, buddy." Davenport was probably one of those guys who call everybody "buddy" whose name they can't remember. "Phil Donaldson and his wife have to go to Hawaii on Saturday. So how would you and your buddy like to sit in my Sky Box at the Coliseum and watch the game?"

"We'd love to!" Pete was obviously elated.

"Terrific!" Davenport was delighted to see his largesse so warmly received. "Go to the will-call counter at noon, and tell them that Mr. Davenport has left tickets for you. Then say 'gingerbread.'" There were other guests at the party, and Davenport headed off to jaw with them. We put in another couple of hours at the party and returned home. Needless to say, Davenport's unexpected bonus had made a big hit with Pete, so much so that the next day he boosted his action to three hundred on the Raiders minus four.

Saturday morning dawned clear and bright, and we headed over to the Coliseum in plenty of time to get there for a noon will-call, to say nothing of a one p.m. kickoff. The burning question of the day was whether Bobby Joe Whitney, the Raiders' aging star quarterback, his backup Dan Driscoll, or the rookie flash Mike Stankowitz would start at quarterback.

The Raiders had stumbled at the start of the season. The word was that Bobby Joe was injured but wouldn't admit it, and Driscoll didn't seem to fit into the Raiders scheme. Midway through the season, the silver-and-black had posted a lackluster four and four record, and the betting was that they'd be out of the playoffs for the first time in five years. With the L.A. media on his back, the coach had made a brave decision: He decided to start Stankowitz for the ninth game of the season, a home game against a perennial doormat. Stankowitz had been signed out of Washington State for a multimillion-dollar contract. Along with Stankowitz had come Hugh Dryden, Stankowitz's former high school coach, who early in Stankowitz's career had sensed that the big guy was a potential meal ticket. Stankowitz had turned down offers from numerous football powerhouses because Washington State had allowed Dryden to join the staff as an assistant coach, and Stankowitz had made the same deal for Dryden with the Raiders.

The Raiders had never looked back. In his first game, Stankowitz had thrown for three hundred fifty yards, four touchdowns, and no interceptions. The Raiders had gone six and two in the last half of the season to finish with a record good enough to host the playoff game between two wildcard teams: the Raiders and the Pittsburgh Steelers.

The magic word "gingerbread" got us to Davenport's Sky Box, where our host greeted us with his traditional, "Hi ya, buddy!" He led us to a couple of extremely comfortable chairs and handed each of us a pair of binoculars (the field is some distance away from the Sky Boxes, as you can well imagine).

Evidently, Davenport favored the hands-on style of management, for the Sky Box next door housed a PC and several seats by the big plate glass window for the coaching section that constituted the Eye in the Sky. It was currently inhabited by Tom Woodhouse, special teams coach, and the aforementioned Hugh Dryden. The two of them reminded me of Mission Control at NASA, as they were equipped with headsets, playbooks, and an armful of multicolored charts. It was clear from the hustle and bustle that the two coaches were in "do not disturb" mode.

Pete watched the pregame preparations through his binoculars. Before I knew it, it was 12:45, fifteen minutes to kickoff. The

Raiders trotted back on the field. Pete was still glued to the binoculars. Then I noticed that there was a general air of unease circulating through the Sky Box.

"That's funny," Davenport remarked. "Where's Stankowitz?"

The other members of the residential portion of the Sky Box took up their binoculars. Nobody seemed to know. Woodhouse and Dryden were certainly able to overhear Davenport's remark, but evidently they were preoccupied with planning strategy, for they seemed to take no notice of the proceedings.

"Tom!" Davenport snapped at Woodhouse. Woodhouse came to attention, as no doubt would I when addressed in an imperious fashion by the source of my daily bread.

"Yes, Mr. Davenport?"

"Call the coach and ask him where Stankowitz is!"

Woodhouse complied. A short discussion followed. "Nobody seems to have seen him since the team left the locker room."

"Well, tell somebody to get back to the locker room and find him real quick. And if he isn't found and we win the coin toss, elect to kick off."

"Right away, Mr. Davenport."

By now the crowd was abuzz. The big clock by the peristyle end of the Coliseum showed a couple of minutes before one. The cocaptains walked out to the center of the field to enact the coin-toss ritual.

A cry came from Woodhouse. "They've found him. He's in the locker room, lying on the floor, passed out. Unconscious or something. There's half a banana lying by his side."

Pete had put down his binoculars and was paying close attention. Since I was clearly at sea, he said, "Stankowitz always eats three bananas before the start of a game. Bananas are a great source of potassium and carbohydrates."

Well, at least Pete was on the case, even though at that time I don't think either of us knew it was a case. The Steelers punted after going three and out in their first series of downs. The Raiders had the ball at their own thirty-five, and Bobby Joe Whitney walked onto the field to direct the Raiders' efforts.

"Wonderful!" Pete muttered disgustedly.

"What's the problem?" I asked.

"Whitney's not in shape. He hasn't played for almost two months, except in garbage time, and he's got bruised ribs. I don't like my chances."

By now, I had hung around Pete and other bettors long enough to learn that betting on a team entitled the bettor to temporary possession of that team, even if that possession existed only in the mind of the bettor. The Raiders, who were four-point favorites, played a gallant first half and left the field tied at ten apiece. By now, reports were being relayed almost continuously on Stankowitz's condition.

Those reports were extremely disquieting. The banana had been sent out to a lab for analysis and was shown to have been drugged. Since there was evidently a lot of fruit in the locker room for players to eat, I wondered how the individual who drugged the bananas could have been sure that Stankowitz, and not a third-string offensive tackle, would have eaten them. That was cleared up by a statement from the trainer, who said that Stankowitz favored bananas that were only partially ripe. He usually came in early and chose three, which were then set aside. So anyone could have drugged them.

Even though Stankowitz couldn't come back to lead the team, it was a tremendously exciting game. The two teams were never more than a touchdown apart. The Raiders, trailing by a point with a minute and a half to play, took over at their own eight-yard line and mounted a last-ditch desperation drive. Whitney fought the clock, the Steelers' defense, and his own injuries to a standstill, showing flashes of the brilliance that had made him the Raiders' number one for the better part of a decade. Finally, with three seconds to go, the ball was positioned for a forty-five yard field goal try. With the season hanging in the balance, the ball hit the crossbar and bounced through. Raiders 19, Steelers 17.

Needless to say, there was jubilation in the Sky Box. After all the tumult and the shouting had died down, I saw Pete pull Davenport aside and whisper a few words in his ear. At first, Davenport looked perplexed, but then he appeared to understand.

We retrieved our car and headed home. Pete was not a happy camper. I tried to cheer him up.

"Come on, Pete, you've lost bets before. Three hundred bucks isn't the end of the world."

"Five fifty," he muttered through clenched teeth.

"Huh?"

"Three hundred on the Raiders. Two hundred on over thirty-seven and a half. And fifty vig. Not only that, I analyzed it right. I *know* Stankowitz would have made the difference. Bobby Joe was so banged up, he could only throw short passes."

/ (See An Introduction to Sports Betting for an explanation of vig on p. 233.)

As we got back to the house, a thought occurred to me. "By the way, Pete, I saw you pull Davenport aside at the end of the game for a little chit-chat. What were you talking about?"

"I just had an idea. Maybe it wasn't very important." Pete seemed strangely disinclined to discuss it, so I switched to another topic of conversation.

I had some time to myself the next day, so I engaged in a little game of "Who Doped the Banana?" which had already risen to the top of the charts as L.A.'s favorite quiz show. The leading candidate, according to the papers, was the trainer, who was obviously in closest touch with the eating habits of the players. Bobby Joe Whitney also fell under suspicion, as he was playing out his option year, and his value would decrease substantially in next year's market unless he could show that he still had the right stuff. There was a tie for third among everyone else but Stankowitz, as astronomical sums of money could be made betting on a game in which the star quarterback would be unable to play. Pete still wouldn't discuss it.

The next day was New Year's Eve. At about ten in the morning, the doorbell rang. It was a messenger bearing an envelope addressed to Pete. Pete signed for it, lighting up like a kid on Christmas morning. He opened the envelope. The first item I saw was a season ticket to the next year's Raiders games, exhibition games and playoffs included—also parking, which at an L.A. sports event is sometimes just as valuable as the tickets. The second item was a note imprinted "From the Desk of Harvey Davenport," which said, "You were right, buddy." It was the third item that won the "most favored item" prize. It was a check made out to the firm of Lennox and Carmichael for $10,000!

It didn't require too much brainwork to realize that Pete had tipped Davenport off to the individual responsible for doping the bananas. What else could be worth ten thousand bucks?

"So that's what you were whispering to Davenport at the end of the game," I remarked.

"Yeah." Pete was being unusually laconic, even for Pete.

"Well, I give up," I said. "I've thought about it quite a bit. So, apparently, has everyone else in L.A. Who did it?"

"I couldn't be sure, but I thought that Hugh Dryden was the most likely candidate."

"Dryden? But he wouldn't even have that job if it weren't for Stankowitz! As a matter of fact, he's ridden on Stankowitz's back ever since he was his high school coach!"

"Exactly my point." Pete poured himself a cup of coffee. "Do you remember when we first noticed that Stankowitz wasn't on the field? What did Dryden do?"

I thought back. "Just what he was doing before. He checked over the game plan with Woodhouse."

"That's what I didn't understand. Here's Stankowitz, Dryden's meal ticket, not showing up for the most important game in his career. Wouldn't you think that Dryden would have been nearly beside himself with worry? Instead, he carried on as if it was business as usual."

I looked at him with surprise, liberally mixed with admiration. "Pete! Don't tell me you're becoming a student of human nature!"

He almost blushed. "Well, I couldn't be sure. That's why I told Davenport that he should get a team of men to watch Dryden. There was a chance that he'd give himself away, either by trying to cover up something or by collecting his payoff."

I took the check and went to the bank, making sure that it got there before it closed. Banks close early on New Year's Eve, and I always like checks to clear as soon as possible—even if they're made out by people as wealthy as Harvey Davenport. My conscience was twitching, and I never like that. On the other hand, Pete and I are a partnership, and sometimes I do much more than 50% of the work. It was a problem, but I finally solved it to my satisfaction. I deposited $9,450 in our account, and took the remaining $550 to give to Pete to cover his associated gambling losses, deciding that, in this case, it was probably a justifiable business expense. After all, I reasoned, if Pete hadn't programmed the Traveling Salesman Problem to get us to arrive at Harvey Davenport's party just when Davenport had a couple of extra tickets, we would have been out ten thousand bucks.

I hadn't really felt like going out this New Year's Eve. Sometimes I get pangs of loneliness around holidays, and I was feeling a little blue. Pete and Julie Rydecki had recently become an item and had made reservations at one of the trendier nightclub-restaurants, but Julie had come down with the latest incarnation of Asian flu and had canceled.

Pangs of loneliness are often exacerbated by the wrong company, and I just didn't want to go to a party with lots of couples. Pete didn't either, so we decided to throw a couple of steaks on the indoor grill and simply toast the New Year. There's always a football game on New Year's Eve, and as I had nothing to do and didn't want to go back to an empty guesthouse, I watched it with Pete. The episode of *The Proud and the Passionate* in which Julie had a bit part was due to rerun a little later, and Pete had obviously planned to watch it with Julie until she came down with Asian flu. You may recall that not only hadn't we seen that episode the first time it showed, but we hadn't recorded it either, so this gave Pete another chance. After the episode, he got on the phone with Julie about half an hour before midnight, and it was my cue to exit. I wished him Happy New Year, and headed back to the guesthouse.

I switched on the light.

"Happy New Year, Freddy."

I had a hard time believing my ears, but my eyes were there to visually confirm that it was Lisa! And not merely occupying space, but wearing the off-one-shoulder black dress that always took my breath away every time I saw her in it.

"It's certainly starting out that way." I did have enough breath for that remark. Questions were vying with each other for voice time, and the first two to emerge were "How did you get in? Did I leave the door unlocked?"

"No, Pete sent me a key."

My jaw dropped. "He *what*?"

"He sent me a key. I called him up and told him that our firm was opening an office in Los Angeles, and I had been asked to head it. That was the job opportunity I told you about that I didn't want to jinx. I said that I wanted to surprise you for New Year's—that is, if you wanted to be surprised. Pete told me that if there had been a line posted in Vegas on whether or not you wanted to be surprised, they would have taken it off the board. I took that as a yes."

I didn't know whether Pete's assessment of my feelings on this matter had taken place before or after he told Harvey Davenport that he should keep a close eye on Hugh Dryden, but maybe Pete really was becoming a student of human nature.

Lisa paused for a moment. "It's a little chilly in here." Possibly her choice of outfit had something to do with that statement. "Maybe you could light the fire."

I went over to the old-fashioned fireplace. Lisa had arranged the logs and kindling carefully above paper to be used as tinder. Detectives are trained to be observant, and I saw that Lisa hadn't used the available old newspaper for that purpose, so I looked a little more closely at the paper. It was our separation agreement!

"Good choice of tinder," I said approvingly.

I made sure that the flue was open, as this was clearly one of those moments that would have been ruined by the buzzing of a smoke alarm, and lit one of the long matches. We watched the paper burn, and then the kindling. Just as the logs were beginning to catch fire, the clock struck midnight. I could hear sirens, horns honking, and—as happens every New Year's, celebratory gunshots from the general direction of Venice, which borders Santa Monica.

"Happy New Year, Lisa."

We kissed—and not just a Happy New Year kiss. When we finally broke apart, Lisa asked, "Do you think we should go over and wish Pete a Happy New Year?"

"It can wait until morning," I replied, as I put my arms around her.

APPENDIXES

CONTINUING THE INVESTIGATIONS

APPENDIX 1

MATHEMATICAL LOGIC IN "A CHANGE OF SCENE"

Before you dive into this section, let me repeat something I said earlier—you don't have to read this! I think a fair amount of learning takes place by erosion—if you're simply exposed to something often enough, it will sink in. Maybe not deeply, but enough to give you the idea. Hang around with musicians, you get some idea of what goes into music—maybe not enough to make you a musician yourself, but you'll be a lot more knowledgeable about it than if you spent no time on it at all.

However, I hope that you'll try reading some of these sections. They're not deep, and if you get fed up, you can always go on to the next story.

Sherlock Holmes was fond of telling Watson that when you eliminate the impossible, whatever remains, no matter how improbable, must be the truth. It's certainly a simple, insightful, and elegantly phrased remark. However, it does not come as a stunning surprise because most of us are already aware of the inherent logic behind Holmes's statement.

It is perhaps fitting that Sherlock Holmes, for whom logic was a sine qua non, was an Englishman, for it was another Englishman, George Boole, who was most responsible for the invention of symbolic logic.

Before George Boole, mathematicians concerned themselves with mathematical objects such as numbers and geometric figures. A goal of mathematics then, as now, was to prove theorems—about numbers, geometric figures, and such. To George Boole goes the honor of being the first extensive investigator of the nature of proof (not to slight the Greek philosophers, who made the initial contributions in this area).

One of the reasons that mathematics is so successful is its relentless focus on concepts whose definitions are unambiguous. Boole focused his attention on statements or sentences that were either unambiguously true or unambiguously false. Such statements are called propositions. Throughout appendix 1, the letters P, Q, and R are used to denote propositions, and the letters T and F are used as abbreviations for "true" and "false," respectively.

To each proposition there is an opposite, which is called NOT P. A proposition can be negated simply by sticking the phrase "It is false that . . ." in front of the proposition. For instance, if P is the proposition "Today is Thursday," the opposite of P (sometimes called the negation of P) is the proposition "It is false that today is Thursday."

From simple propositions, more complicated ones can be built up through the use of the logical terms OR and AND, which behave pretty much the way they do in ordinary English. The proposition P AND Q, for instance, is true only when both P and Q are true—just as you would expect. If someone said to you, "The New England Patriots won the Super Bowl in 2015, and Austin is the capital of Texas," you'd undoubtedly agree that the statement was true, although maybe you'd need to do a little Googling first. However, if they said, "The New England Patriots won the Super Bowl in 2015, and $2 + 2 = 5$," that would elicit a "whoa, there" on your part.

The term OR is a little more subtle because we use the word "or" in English in two different ways. The exclusive "or" is used to preclude one of the two possibilities, as in "Did you register as a Democrat or as something else?" You can't do both—at least, not legally. However, the inclusive "or" allows both possibilities to be selected. When the waiter at a restaurant asks you if you would like coffee or dessert, he obviously won't be offended if you choose to have both—especially as the size of his tip will probably increase. At some time in the past, mathematicians decided to use the inclusive "or," just as they decided to use the symbol + for addition rather than something else—and so we'll go with that.

This information can be quickly summarized in tabular form. The layout that follows is known as a truth table.

P	Q	P OR Q	P AND Q
T	T	T	T
T	F	T	F
F	T	T	F
F	F	F	F

With the four rows, we have covered all the possible true-false combinations for the two propositions P and Q, and the other columns give the truth values of the proposition at the top of the column for the truth values of P and Q in the same row.

Here's a very informative, very short truth table.

P	NOT P	P OR NOT P	P AND NOT P
T	F	T	F
F	T	T	F

In words, for any proposition P, the proposition P OR NOT P is always true, and the proposition P AND NOT P is always false. Propositions that are always true are known as *tautologies*.

Some truth tables are important but not especially interesting. It is easy to show that P OR Q and Q OR P have the same truth table. This isn't surprising, as if your waiter asks, "Will you have dessert or coffee?" it's the same question as if he asked, "Will you have coffee or dessert?" Similarly, P AND Q and Q AND P have the same truth table.

It is possible to use parentheses to construct ever more complicated propositions, much as parentheses are used in arithmetic and algebra for exactly the same purpose. If P, Q, and R are propositions, we can construct the compound proposition P OR (Q AND R) by first constructing the proposition Q AND R and then taking that proposition and OR-ing (as the computer folk are fond of saying) the proposition P with the proposition Q AND R.

We can compute the truth value of a complicated proposition from the truth values of its components simply by stripping away levels of parentheses. Just as we compute the numerical expression $(2 + 3) \times (3 \times (4 + 7))$ by working from the inside out, we can do the same thing with complex propositions.

Arithmetically, $(2 + 3) \times (3 \times (4 + 7)) = 5 \times (3 \times 11) = 5 \times 33 = 165$. To see how this works in symbolic logic, we'll assume that P is true, Q is false, and R is false. To compute the truth value of the following logical expression, (P OR NOT Q) AND (NOT P OR R), we just replace each proposition by T or F as we compute it.

1) (P OR NOT Q) AND (NOT P OR R)
2) (T OR NOT F) AND (NOT T OR F)
3) (T OR T) AND (F OR F)
4) T AND F
5) F

Now let's return to Sherlock Holmes. How can we analyze his remark that, when you have eliminated the impossible, whatever remains, however improbable, must be true?

Let's suppose that P and Q are propositions such that P OR Q is true. Suppose further that Q is false. How can we conclude that P must be true?

One look at the truth table for P OR Q should make it fairly obvious.

Line	P	Q	P OR Q
(1)	T	T	T
(2)	T	F	T
(3)	F	T	T
(4)	F	F	F

Since P OR Q is true, line (4) is eliminated. Since Q is false, lines (1) and (3) are likewise out. No matter how improbable P may be, it must be true, as line (2) is the only one remaining, and P is true in line (2).

Now things get a little complicated. Boole decided to assess the validity of the argument IF P THEN Q on the basis of the true-false values of the propositions P and Q. What Boole decided was that the important thing was to make sure that any argument which started with a true premise (the premise is the P in IF P THEN Q) and ended with a false conclusion (the conclusion is the Q in IF P THEN Q) would be labeled as false. After all, if you start with the truth and reach a false conclusion, your argument must be fallacious. To single out these fallacious arguments, Boole made all other IF P THEN Q statements true, by fiat.

This resulted in the following truth table for IF P THEN Q.

P	Q	IF P THEN Q
T	T	T
T	F	F
F	T	T
F	F	T

Let's look at the compound proposition IF ((P OR Q) AND NOT Q) THEN P. So that everything will fit on one line, let R denote the proposition (P OR Q) AND NOT Q.

P	Q	NOT Q	P OR Q	(P OR Q) AND NOT Q	IF R THEN P
T	T	F	T	F	T
T	F	T	T	T	T
F	T	F	T	F	T
F	F	T	F	F	T

No matter what the truth values of P and Q, the proposition

IF ((P OR Q) AND NOT Q) THEN P

is always true!

Admittedly, when Sherlock Holmes used it, he assumed implicitly that either P or Q is true. Nonetheless, no matter what the truth values of P and Q, IF ((P OR Q) AND NOT Q) THEN P must be true.

When two propositions have the same truth table, they are said to be *logically equivalent*, and we often use those equivalences in everyday speech. When the waiter asks you if you would like coffee or dessert and you tell him, "No," he knows that you do not want coffee and you do not want dessert. You—and the waiter—have used the fact that NOT (P OR Q) is logically equivalent to (NOT P) AND (NOT Q), and Pete used it in the story to conclude that the contact was not going to meet Hazlitt and was also not going to meet Burns.

Boolean logic, as this branch of mathematics is known, has gone far beyond what Boole could ever have imagined. Not only is your computer constructed on its principles, but also every time you do an advanced search with a search engine, you are using Boolean logic as well.

APPENDIX 2

PERCENTAGES IN "THE CASE OF THE VANISHING GREENBACKS"

As Pete observes in the story, percentages are a source of substantial confusion. Most of this confusion comes from a mistaken belief that percentages work the same way as numbers, with a gain of 20% compensating for a loss of 20%. As we saw in the story, a gain of 20% does not compensate for a loss of 20% because the 20% gain is not figured using the same amount as the 20% loss.

A knowledge of basic algebra can be quite helpful in eliminating much of the confusion surrounding percentages. There's a little algebra in the material that follows but hopefully not enough to cause you sleepless nights.

Innumeracy, as Pete points out in the story, is the arithmetic equivalent of illiteracy. Those who have succumbed to illiteracy, however, realize that they cannot read. They know that it can profoundly affect their lives and often take steps to remedy this problem.

Innumeracy is much more insidious than illiteracy. The victims often do not realize that they are innumerate. Illiteracy is condemned, but innumeracy is not regarded in the same light. There are those who feel that such attributes as artistic creativity go hand in hand with innumeracy—indeed, there are even some who proudly flaunt their innumeracy.

This is a great pity because, like any disease, the ripple effects of innumeracy spread throughout our society. It is probably not an exaggeration to say that elimination of innumeracy would save our society tens, and perhaps hundreds, of billions of dollars annually.

Any study of innumeracy would undoubtedly find that confusion concerning percentages is a contributing factor. The term "per cent" is an abbreviation of the Latin phrase "per centum," which means "each hundred." Thus 3% means 3 for each hundred; 3% of 100 is 3, and 3% of 400 is 12.

COMPUTING PERCENTAGES

To find a percentage of a given number, you just need to multiply the number by the percentage, and divide by 100.

Example 1: To find 7½% of $350, multiply 7½ by $350, obtaining $2,625, and then divide by 100 to get $26.25.

Computing percentages is straightforward, and most people do not have too much difficulty doing so. Finding the cost of an item on which the sales tax is known is a little more difficult and requires setting up and solving a simple equation.

Suppose we want to find the purchase price of an item on which a 7% sales tax came to $11.20. Let P denote the purchase price. Then 7% of P is $7\ P/100 = 0.07\ P$. Therefore,

$$0.07\ P = \$11.20$$
$$P = \$11.20/0.07 = \$160$$

You can check that $160 is the right price simply by taking 7% of $160 and observing that it equals $11.20.

Example 2: Rutabaga Preferred stock rose 15% last year. If the stock went up 13½ points, at what price did it start the year?

(Note: A "point" is actually investor-speak for a dollar. A stock selling for 50 points is selling for $50 for one share of stock.)

Solution: If S is the starting price, 15% of S is 0.15 S, so

$$0.15\ S = 13.5$$
$$S = 13.5/0.15 = 90$$

The stock started at 90. Checking, 15% of 90 is 13.5. ∎

From here we move on to markups and markdowns. Markups generally don't cause problems mathematically, but since many of them come from various governmental institutions imposing taxes, they're a major annoyance. In real life, that is—but the major mathematical annoyance comes from confusion with discounts. Check out this common mistake, which occurs because many people erroneously compute the original price, from which a discount is taken, by adding that percentage to the discounted price

Example 3: After a 20% discount, a TV sells for $120. What was the original price?

Solution: The mistake is to take 20% of the discounted price of $120, which is $24, and add that to the discounted price of $120, arriving at an erroneous original price of $120 + $24 = $144.

You, of course, now know better. Let S denote the original price of the TV. Then 20% of S is 0.20 S, and so

$$S - 0.20\ S = \$120$$
$$0.8\ S = \$120$$
$$S = \$120/0.8 = \$150$$

Checking, 20% of $150 is $0.20 \times \$150 = \$30$, and when $30 is subtracted from $150, the result is $120. ∎

Now let's take a look at the trap into which the city of Linda Vista fell. Why can't we reduce an individual's taxes by 20% if the tax base increases by 20%? If there are originally T taxpayers, and each taxpayer is assessed D dollars, then the total revenue is clearly TD dollars. A 20% increase in the number of taxpayers will add 0.20 T taxpayers to the original T taxpayers, so there will now be T + 0.20 T = 1.2 T taxpayers. A 20% reduction in the taxes assessed each taxpayer will be 0.20 D dollars, so each taxpayer will now pay D - 0.20 D = 0.80 D dollars. Therefore, the total revenue will be the number of taxpayers multiplied by the tax per taxpayer. This comes to 1.2 T × .8 D = 0.96 TD dollars. This is only 96% of the original revenue.

/ (Percentage calculation continued from p. 18)

Incidentally, did you catch on to the fact that Pete was able to compute the number of taxpayers simply from the information that taxes had been reduced from $100 to $80 and that the city of Linda Vista was $396,000 short? Once he found out that everybody assessed had paid up, Pete hypothesized that the city council had fallen into the classic innumeracy trap of thinking that a 20% gain in the number of taxpayers compensated for a 20% loss in revenue per taxpayer. In that case, as we discovered above, the shortfall would have been 4%, since the revenue was only 96% of the original revenue. If the total revenue is denoted by R, then 4% of R would be $396,000. Therefore,

$$0.04 \ R = \$396,000$$
$$R = \$396,000/0.04 = \$9,900,000$$

At $100 per taxpayer, the number of taxpayers in the previous census would have been $9,900,000/100 = 99,000 taxpayers.

Innumeracy in this respect can have potentially catastrophic consequences. A doctor may tell a nurse to reduce the dosage of a drug by 50%. When the patient relapses, the doctor tells the nurse to raise the dosage by 50%. Disaster! The doctor may think the patient is receiving the same amount of medication as he or she did originally, but the patient is only receiving three-quarters of the original amount. One shudders at the thought of a similar error being made with an airplane whose fuel has been depleted by 50%.

Many of the misunderstandings concerning percentages occur because of a failure to realize that the computation of a percentage requires a base number upon which one computes the percentage. Suppose that a stock is selling for 100, and the price rises 30% and then declines 30%. The base number for figuring the price *rise* percentage is 100; 30% of 100 is 30, so the price after the rise is 130. This number, 130, is the base number for figuring the price *decline* of 30%. And 30% of 130 is 39, so the stock price falls to 130 – 39 = 91. Notice that, even if the stock had fallen 30% first and then risen 30%, the final price would again be 91. That's because we are performing two successive multiplications, and it doesn't matter in which order we perform them.

Failure to understand percentages has repercussions in other areas of mathematics. Percentages are often used to convey probabilistic notions. For instance, a meteorologist might say that there is a 50% chance of rain on Saturday and a 50% chance of rain on Sunday. A recent poll showed that many people were under the impression that the above forecast was equivalent to saying that it was certain that it would rain during the weekend! Well, it's not, and we shall have more to say about this in chapter 9.

APPENDIX 3

AVERAGES AND RATES
IN "A MATTER OF TIME"

If mathematicians were to vote on the most useful notion in mathematics, the concept of averages would be up close to the top. In fact, many would probably award it the title. Averages occur throughout all of mathematics. They represent one of the best ways of summarizing past information, and in the absence of more pertinent data, the best way to predict the future. Averages play critical roles in such widely diverse topics as percentages, probability and statistics, algebra, and calculus.

An average is a quotient, and one model for division is sharing a quantity of items in a fair fashion. If four people are to share twelve slices of pizza fairly, how many slices should each person receive?

Simple as this example may seem, it can be used to provide an easy introduction to the concept of an average. If twelve slices of pizza are shared among four people, the average number of slices each person receives is three. Of course, this does *not* mean that each person actually receives three slices. An average in this instance represents a way of summarizing data by looking at what would have happened if fair sharing had actually taken place.

When we say that the average is three slices, there is a very important, but often unspoken phrase: "per person." An average is a quotient, and a quotient consists of a numerator and a denominator. When we are dealing with real-world quantities, the numerators and denominators are measured in units. The numerator units in the above example are slices of pizza, the denominator units are persons. To fully understand an average, one must know what is being shared (the numerator units—pizza slices) and among what the shared quantity is being shared (the denominator units—people). The units of measurement for averages are "numerator units per denominator unit"—in this instance, slices per person. All the concepts in this chapter involve quotients.

THE IMPORTANCE OF UNITS

When computing an average, a number by itself is meaningless—both the numerator and denominator units must be specified. To see how important this is, ask yourself if you would take a job if the salary was 5.

Assuming that the job isn't distasteful or dangerous, you almost certainly would take the job if the salary was $5 per second. You almost certainly wouldn't take the job if the salary was 5 cents per year.

Averages: Summarizing the Past, Predicting the Future

An average is a quotient. Baseball, the national pastime (although rapidly being eclipsed by football), provides an excellent source for the computation of averages. A player's batting average is the quotient of the number of hits the player has achieved divided by the total number of official at-bats (an official at-bat occurs any time a player is not automatically awarded first base via a base on balls or being hit by a pitch). If a player has 500 official at-bats and gets hits in 150 of these, his batting average is $150/500 = 0.3 = 0.300$ (pronounced "three hundred"). (A player whose batting average rounded to the nearest thousandth is .273 is said to be hitting "two seventy-three.") A player's batting average is the average number of hits per official at-bat.

Computing averages is not difficult, but it is surprisingly easy to be misled by the form in which the information is presented.

Example 1: Evan buys $6.00 worth of hamburger at $1.50 a pound and then goes to another store where he buys another $6.00 worth of hamburger at $1.00 a pound. What is the average price of the hamburger?

Solution: If you computed the answer by saying that, since the same amount was purchased, the average price of the hamburger must be $1.25, the average of the prices $1.50 and $1.00, you have fallen into exactly the same trap as Freddy did during the story!

Many problems involving averages simply require you to keep focused on the numerator and denominator of the quotient that you use to compute the average. In this case, the numerator is $12.00, the amount of money spent, and the denominator is 10 pounds, the amount of hamburger purchased. The average price is therefore $12.00/10 pounds = $1.20 per pound. ∎

Many of the mistakes made in computing averages are variations of the error that Freddy made during the story. Suppose that one is given two different data sets and computes an average for each data set, and then computes the average of all the data together. In general, it is *not* true that the average for all the data is the average of the averages for each data set. In example 1, the trap is to compute the average of two average prices. The way to avoid this trap is to compute the numerator and denominator for the entire data set—computing the average of two averages is very likely to lead to a wrong answer.

/ (Calculating averages continued from p. 27)

In the story, Freddy made the common mistake of assuming that the average rate is the average of the rates. It certainly sounds convincing, but it is only true when the denominators of both rates are equal. The actual average rate of Freddy's round trip to San Diego can be computed by looking at the total distance traveled (240 miles) and dividing by the total time taken (3 hours to go the 120 miles to San Diego at 40 miles per hour; 6 hours to return at 20 miles per hour). Although the average of 40 and 20 is 30, the average speed on the trip is $240/9 = 26\frac{2}{3}$ miles per hour. As Pete observed, the actual average speed is different from Freddy's estimate of 30 miles per hour because time, the denominators of the two rates, differed for the two legs of the round trip.

That's not the only way that averages can cause confusion, as the following example shows.

Example 2: The four executives at Mirage Financial are paid annual salaries of $100,000, and the six assistants get annual salaries of $40,000. The executives got raises of $10,000, and the staff got raises of $2,000. The company told its stockholders that the average raise was 7%, and it told the employees that the average raise was 8.125%. What's going on here?

Solution: Welcome to the wonderful world of creative accounting! There were four raises of 10% and six raises of 5%; the average of these ten percentages is 7%. On the other hand, the initial payroll was $640,000, and after the raises the payroll was $692,000, an increase of 8.125%. ■

There is some truth in both of these numbers—they are both correctly figured, but they are differently defined. This example indicates why it is often difficult to find out what is really happening with the finances of a company that has a complicated balance sheet.

Averages are often used as a basis for estimates of future performance. This reliance on averages to estimate future performance exists even where the reasons that resulted in that average are not well understood. If a hundred years of weather data show that the average rainfall in an area is thirty inches per year, planning for water use and distribution is made on the assumption that the future rainfall will continue to average thirty inches per year.

Using past averages as future estimates is common practice. A company estimates future sales by looking at past averages. It is important to realize that averages can be used to estimate the future, and plans based on those estimates may often be extremely useful, but averages based on past data cannot predict the future. Even though past rainfall may have averaged thirty inches per year, the coming year may see either a drought or a flood.

Rates

Rates are quite similar to averages, as both averages and rates are expressed as quotients.

From a mathematical standpoint, there is no difference between an average and a rate. However, there is a sort of unspoken agreement that the word "average" is generally used to summarize data, whereas the word "rate" is generally used to describe an ongoing process or to facilitate exchanges.

If twelve slices of pizza were eaten by four people, each person ate an average of three slices; this summarizes what happened in the past. We could also say that the rate of pizza consumption is three slices per person; here the ongoing process of eating pizza is being emphasized. The numerical measures (three) of both average and rate are the same, the units of measurement (slices per person) are the same, so from the mathematical standpoint, they are indistinguishable.

Exchanging money for goods or services is basic to business. If hamburger costs $2 per pound, a denominator unit (pound of hamburger) can be exchanged for $2. This exchange is how business commonly sees rates, but exchanges also underlie rates in other situations. For example, if a car goes at a rate of forty miles per hour, a denominator unit (one hour) can be exchanged for forty miles. This may be a somewhat strange way to view driving speed, but it shows that the rate concept has applications in different areas.

SEQUENCES AND ARITHMETIC PROGRESSIONS IN "THE WORST FORTY DAYS SINCE THE FLOOD"

The ability to recognize and use patterns is one of the most important aspects of intelligence. It is only in the last ten thousand or so years that we have created societies and the advances that go with them. Props to us! I'm pretty sure these advances began when we first started to use the pattern of recurring seasons to develop agriculture.

Predictions based on patterns form the foundation of science and technology, as well as everyday life. Medical researchers study the pattern of the beating heart in the hopes of isolating cues that will alert them to incipient heart attacks, but you and I also try to pay attention to the patterns of behavior of our friends, our co-workers, and our loved ones. Things go better when we figure out how people will react.

One of the great things about numbers is that they provide an environment that makes it possible to recognize patterns, and these number patterns often reflect things that happen in the real world.

SEQUENCES AND ARITHMETIC PROGRESSIONS

A *sequence* is an unending string of numbers. Some well-known examples of sequences are

The counting numbers: 1, 2, 3, 4, . . . (the dots indicate that the numbers continue);

The odd numbers: 1, 3, 5, 7, . . . ; and

The prime numbers: 2, 3, 5, 7, . . . (a prime number has only two whole number divisors, itself and 1. So 13 is prime because its only whole number divisors are 13 and 1, and 12 is not prime because it has 2, 3, 4, and 6 as whole number divisors).

The numbers that make up a sequence are called its *terms*. In the sequence 1, 3, 5, 7, . . . of odd numbers, 1 is called the first term of the sequence, 3, the second term, 5, the third term, etc.

Letters can be used in algebra when we wish to talk about numbers and their properties without specifying a particular number. For example, we use the letter x in the equation $2x + 6 = 2(x + 3)$ to denote any number. When we wish to talk about sequences or properties of sequences without specifying a particular sequence, we use the notation $a_1, a_2, a_3, \ldots$ to denote a sequence. The numbers $1, 2, 3, \ldots$ in the above notation are called *subscripts*. For instance, a_7, which is read "a sub seven," or more casually, "a seven," is the seventh term of the sequence. When we wish to talk about a term of the sequence without specifying a particular term, we use the notation a_n.

There are several different ways to describe sequences. The sequence 1, 3, 5, 7, . . . of odd numbers can be described

(1) by using words: the nth term of the sequence is the nth odd number.

(2) by using a formula: $a_n = 2n - 1$. Notice that using a formula enables us to compute the exact value of *any* term in the sequence. For instance, $a_{100} = 2 \times 100 - 1 = 199$.

(3) by using a *recursive definition*, which describes a sequence in roughly the same way that one would give instructions on how to use a ladder: put your foot on the first step, and whenever you are standing on a rung, put one foot on the

next higher step and bring your other foot up to join it. In this instance, the recursive definition would be $a_1 = 1$, $a_n = a_{n-1} + 2$.

We start with the definition that $a_1 = 1$. When we let $n = 2$, the recursive formula becomes

$$a_2 = a_{2-1} + 2 = a_1 + 2 = 1 + 2 = 3, \text{ so } a_2 = 3$$

Now we can let $n = 3$, and the recursive formula becomes

$$a_3 = a_{3-1} + 2 = a_2 + 2 = 3 + 2 = 5$$

We can now use a_3 to help us compute a_4, etc.

Of the three ways of describing a sequence, describing by means of a formula is the most useful because it enables us to compute directly any term in the sequence. The formula $a_n = 2n - 1$ can be used to compute the one hundredth term in the sequence as we did above. If we wanted to find the one hundredth term using words as the description, we would have to write out the first one hundred odd numbers, and if we wanted to find the one hundredth term by using the recursive definition, we would have to use it ninety-nine times (the first use gave us a_2, the second use gave us a_3, etc.).

Unfortunately, sometimes a formula is not available. In the case of the sequence of prime numbers, there is no known formula or recursive definition that enables us to compute the nth prime number. As a matter of fact, mathematicians have actually been able to prove that it is impossible to find either a formula or a recursive definition to compute prime numbers! And a good thing, too—because this difficulty in computing prime numbers underlies the security that protects your passwords. And, of course, your bank account.

Other examples of sequences that must be defined by words are the daily balances in your checking account or the rolls of a random die.

TWO WELL-KNOWN SEQUENCES

Sequence 1: The even numbers: 2, 4, 6, 8, . . .
   Formula: $a_n = 2n$
   Recursive definition: $a_1 = 2, a_n = a_{n-1} + 2$

Sequence 2: The squares: 1, 4, 9, 16, . . .
  Formula: $a_n = n^2$
  Recursive definition: $a_1 = 1, a_n = a_{n-1} + 2n - 1$

One of the parts of a standard intelligence test is to predict the next number in the sequence. This is a case where too much knowledge is a dangerous thing, at least from the standpoint of scoring well on the intelligence test! For instance, suppose you were asked to predict the next number in a sequence whose first three terms were 1, 2, and 4. If you decide that the sequence is recursively defined by $a_1 = 1, a_n = 2a_{n-1}$, then the next term of the sequence would be 8. If you decide instead that the sequence is recursively defined by $a_1 = 1, a_n = a_{n-1} + n - 1$, then the next term of the sequence is 7!

Sequences can consist of terms other than numbers. This book, for instance, is a sequence of letters, numbers, and symbols (and all the terms after about five hundred thousand are the blank symbol).

ARITHMETIC PROGRESSIONS

Look at the sequence 2, 5, 8, 11, . . . , which we define recursively by $a_1 = 2, a_n = a_{n-1} + 3$. Notice that the difference between any two consecutive terms is 3, which we can see simply by taking the recursion formula $a_n = a_{n-1} + 3$, and subtracting a_{n-1} from each side to obtain $a_n - a_{n-1} = 3$. This is an example of an *arithmetic sequence*, more commonly called an *arithmetic progression*, which is a sequence in which the difference between any two consecutive terms is a constant. This difference is called the *common difference*. In an arithmetic sequence, one can find the common difference by subtracting any term from the term immediately following it. In the sequence 2, 5, 8, 11, . . . , $3 = 5 - 2 = 8 - 5 = 11 - 8$, etc.

If we are given the first term 2 and the common difference 3 of an arithmetic progression, we can immediately write down the recursive definition: $a_1 = 2, a_n = a_{n-1} + 3$. We can also use the pattern to write down the first few terms of the sequence

$$a_1 = 2 = 2 + (0 \times 3)$$
$$a_2 = 2 + 3 = 5 = 2 + (1 \times 3)$$
$$a_3 = (2 + 3) + 3 = 8 = 2 + (2 \times 3)$$
$$a_4 = (2 + 3 + 3) + 3 = 11 = 2 + (3 \times 3)$$

From this pattern, we conclude that

$$a_n = 2 + (n - 1) \times 3$$

Since we could have done exactly the same thing if we had had an arithmetic progression with first term a_1 and common difference d, we have the following two formulas for arithmetic progressions.

FORMULAS FOR ARITHMETIC PROGRESSIONS

An arithmetic progression with first term f and common difference d can be defined recursively by

$$a_1 = f, a_n = a_{n-1} + d$$

or by means of the formula

$$a_n = f + (n - 1)d$$

Arithmetic progressions frequently occur in daily life in the total amount paid by a consumer making installment payments.

Example 1: Susan makes a down payment on a car of $2,000 and monthly payments of $180. Describe her equity (the total of the payments she has made) as an arithmetic progression. If she must make 48 monthly payments, how much will she have paid to buy the car?

Solution: The total amount Susan paid is an arithmetic progression, with $a_1 = \$2,000$ and common difference $180. Therefore, $a_n = \$2,000 + \$180(n - 1)$. After she has made 48 payments, the total paid will be $a_{49} = \$10,640$. ∎

Nature, too, has many examples of arithmetic progressions at its disposal. Many natural phenomena, such as the intervals between eclipses or the closest approaches of planets to the sun (which can be used to mark the changing seasons), represent examples of arithmetic progressions.

Example 2: The great American author Mark Twain (the same one who, as Freddy remarked, found it easy to give up smoking!)

died in 1910, the year that marked the fourth official sighting of Halley's Comet, which appears every seventy-six years. When was Halley's Comet first officially sighted? When did it last appear? When will it next appear?

Solution: We know that $a_4 = 1910 = f + (3 \times 76) = f + 228$, so $f = a_1 = 1682$. (The comet had reappeared many times before then, but this was the year that Halley first recognized it and predicted when it would next reappear.) The last sighting was $a_5 = 1682 + (4 \times 76) = 1986$, and its next appearance will be in $a_6 = 1682 + (5 \times 76) = 2062$.

Incidentally, not only did Mark Twain die during a_4, but he was born during a_3! ∎

SUMS OF ARITHMETIC PROGRESSIONS

/ (Number of cigarettes needed continued from p. 35)
The basic use of multiplication is for repeated addition of the *same* number—3×4 is a shorthand for $4 + 4 + 4$. However, there are certain attractive situations where there are formulas in which multiplication can be used to find the sum of *different* numbers. An example of such a situation was discovered by one of the greatest mathematicians of all time, Carl Friedrich Gauss.

One day while Gauss was attending school, his teacher excused himself from the classroom for a short period and asked the students to add the numbers from 1 through 100 in his absence. Most people, when confronted with this problem, proceed in the obvious fashion. Faced with finding the sum $1 + 2 + 3 + \ldots + 100$, they compute the sum of $1 + 2$, getting 3. To this result they add 3, getting 6. To 6, they then add 4, getting 10. And so on.

Gauss, however, took a different approach. Letting S denote the sum $1 + 2 + 3 + \ldots + 100$, he noticed that S could also be written $100 + 99 + 98 + \ldots + 1$. Writing these two expressions under each other, we have

$$S = 1 + 2 + \ldots + 99 + 100 \qquad [4.1]$$
$$S = 100 + 99 + \ldots + 2 + 1 \qquad [4.2]$$

Adding both these equations gives

$$S + S = (1 + 100) + (2 + 99) + \ldots + (99 + 2) + (100 + 1)$$

The first number in each parenthesis comes from equation [4.1] and the second number from equation [4.2]. But each sum in parentheses adds up to 101, and there are obviously 100 such pairs. So

$$2S = S + S = 100 \times 101 = 10,100$$

and therefore $S = 5,050$.

Gauss was about eight years old when he discovered this phenomenon. In his honor, mathematicians now refer to it as the Gauss trick. It would be the highlight of many a mathematician's career to come up with as cute a trick.

There is obviously nothing special about the number 100. It could have been 1,000, or 8 million, or anything at all. Let's suppose, therefore, that we want to add all the numbers from 1 through N. Letting S denote the sum $1 + 2 + \ldots + N$, we write down the sum forward and backward.

$$S = 1 + 2 + \ldots + (N - 1) + N[\qquad 4.3]$$
$$S = N + (N - 1) + \ldots + 2 + 1 \qquad [4.4]$$

Adding equations [4.3] and [4.4], we get

$$2S = S + S = (1 + N) + (2 + (N - 1)) + \ldots + ((N - 1) + 2) + (N + 1)$$

As before, the first term in each pair of parentheses comes from equation [4.3], and the second term from equation [4.4]. Each parenthetical sum is $N + 1$, and there are obviously N pairs of parentheses. Therefore,

$$2S = S + S = N \times (N + 1)$$

and so $S = (N \times (N + 1))/2$.

Even in an age of ultrahigh-speed computers, it is still extremely important to discover short formulas for sums. After all, why should one make even a high-speed computer add a thousand numbers, when with a little work we can derive a formula that only requires a few additions and multiplications? Sophisticated problems often require billions or even trillions of computations, and shortcut formulas can sometimes eliminate more than 99% of the calculations.

Example 3: In the story, Freddy wished to cut down his cigarettes by one a day. How many cigarettes would he have had to buy if he had wanted to cut down by two a day?

Solution: It's easy to see that $40 + 38 + \ldots + 2 = 2(20 + 19 + \ldots + 1) = 2 \times ((20 \times 21)/2) = 420$. So Freddy would have to buy two cartons and one pack (a carton contains ten packs, and a pack contains twenty cigarettes). Remember that he smokes two packs or forty cigarettes a day. ■

You can find the sum of any arithmetic progression using a similar idea.

APPENDIX 5

ALGEBRA, THE LANGUAGE OF QUANTITATIVE RELATIONSHIPS, IN "THE ACCIDENTAL GUEST"

This material focuses primarily on one of the portions of algebra that causes a lot of trouble—setting up and solving standard story problems. Maybe this describes you. Or, if you are a parent whose child is floundering in this area, read it yourself and impress your child with your skills. If your son or daughter is taking algebra, he or she is almost certainly a teenager, and this will be your last chance to impress him or her for another decade or so.

Algebra is as much a language as it is a subject. Unfortunately, the way algebra is sometimes taught in high school makes it very difficult to see this. In high school algebra, one learns how to simplify algebraic expressions, how to factor polynomials, how to solve quadratic equations, and so on ad infinitum (and, unfortunately, often ad nauseam). What is learned is an assortment of techniques.

Most people who use algebra a lot regard it as a language that describes relationships between quantities. One of the more important aspects of a language is that it enables questions to be raised and answered. In this sense, arithmetic is a language, too—the questions that it raises concern the relationships among numbers. An example of an arithmetic question is, "What is two plus two?" It has only one correct answer and is a question about numbers. As we have mentioned, questions about numbers are the concern of arithmetic.

The extremely funny—and somewhat morbid—cartoonist Gahan Wilson probably spoke for many algebra students in his classic cartoon "Hell's Library." We see a grinning devil with pitchfork, and flames surround the cartoon, leaving no doubt that we are, indeed, in the nether regions. In the background we see a bookshelf. All the books have titles like *The Big Book of Story Problems* and *Even More Story Problems*.

We can't remove all the torments from story problems, but the following two principles provide a great place to start, as they apply to many story problems. Sadly, not all—but many.

THE TOTALS PRINCIPLE AND THE RATE PRINCIPLE

> *Totals Principle*: Totals are sums of subtotals. The Totals Principle is based on addition.
>
> *Rate Principle*: The cost of a number of objects, each of which sells for the same price, is the product of the price times the number of objects. The Rate Principle is based on multiplication and is applicable in any situation involving a rate (not just when the rate is a monetary one).

/ (Algebra continued from p. 48)

Pete uses both of these principles when he expresses the total cost of the car bill in dollars (before taxes) as $60.00 + $0.15 M, where M is the number of miles that the car was driven. The Totals Principle shows that the total cost is the sum of two subtotals; the weekend charge and the mileage cost. The weekend charge is $60, and the mileage cost is given by the Rate Principle as $0.15 M; you're buying M miles at a rate of $0.15 per mile.

All that is left is to solve the equation $60 + $0.15 M = $120.30, and most simple equations yield to repeated applications of the

Golden Rule of Equations; Do unto the right side of the equation as ye do unto the left side. In this case,

$$\$120.30 = \$60.00 + \$0.15 \; M$$
$$\$60.30 = \$0.15 \; M \text{ (subtract \$60 from both sides)}$$
$$402 = M \text{ (divide both sides by \$0.15)}$$

Of course, the \$ sign isn't really necessary here; you could work with 120.30 = 60 + 0.15 M as long as you realized that the eventual answer is to be expressed in miles.

One step up in difficulty level from single linear equations (the word *linear* is used because the graph of $y = 60 + 0.15 \; x$ is a straight line) is a problem with two linear equations in two unknowns.

Example 1: After attending the movies on Wednesday, everyone adjourns to the local pizza palace for pizzas and pitchers. Last Wednesday, they had five pizzas and three pitchers, and the bill came to \$34. This Wednesday, they had six pizzas and four pitchers, for a total of \$42. What is the cost of a pizza? What is the cost of a pitcher?

This problem is different from the ones we have encountered earlier because we are seeking not one, but two items of information. Suppose we let P denote the cost of a pizza, and let B denote the cost of a pitcher (the B is the leading letter of the liquid most frequently filling the pitcher). Last Wednesday, the five pizzas cost $5P$ and the three pitchers cost $3B$, so the total of \$34 results in the equation

$$5P + 3B = 34 \qquad\qquad [5.1]$$

Similarly, the six pizzas cost $6P$, the four pitchers cost $4B$, and since the bill for this came to \$42, we obtain the equation

$$6P + 4B = 42 \qquad\qquad [5.2]$$

To solve this *system of two equations*, there is a standard step-by-step procedure. We pause the solution of this problem to outline it.

PROCEDURE FOR SOLVING TWO EQUATIONS
IN TWO UNKNOWNS

Step 1: Solve one of the equations for one of the unknowns in
terms of the other.
Step 2: Substitute the result of step 1 into the other equation.
Step 3: Solve the resulting equation, which only contains one
unknown.
Step 4: Substitute the result of step 3 into the result of step 1
to obtain the value of the other unknown.

We apply step 1 to equation [5.1] in the pizza–pitcher problem.
The result is

$$P = 6.8 - 0.6B \qquad\qquad [5.3]$$

Substituting this into equation [5.2], we get

$$6(6.8 - 0.6\,B) + 4B = 42$$
$$40.8 - 3.6B + 4B = 42$$
$$0.4B = 1.2$$

So $B = 1.2/0.4 = 3$. Substituting this value into equation [5.3]
yields

$$P = 6.8 - 0.6(3) = 6.8 - 1.8 = 5$$

Although checking is not a formal part of the solution procedure, it should always be performed to be sure that an arithmetic
or algebraic foul-up has not occurred. Assuming that a pizza costs
$5 and a pitcher $3, last Wednesday, the five pizzas cost $25 and
the three pitchers $9, for a total of $34. This Wednesday, the six
pizzas cost $30 and the four pitchers $12, for a total of $42.

Notice that we did not check the problem by seeing whether
the values $B = 3$ and $P = 5$ satisfied equations [5.1] and [5.2]; we
checked the problem by seeing whether the answers we obtained
satisfied the original conditions of the problem. Although it may
seem like these two procedures are the same, they are not—it is
possible to derive incorrect equations from the problem but solve
the equations correctly without solving the problem correctly!

The preceding technique for solving two equations in two unknowns, known as elimination (the elimination takes place in step 3), always works for real-world problems with consistent answers. If, for instance, the price of a pizza was increased from last Wednesday to this Wednesday, the technique will still give answers, but these answers just won't represent actual prices.

The following example uses the Totals Principle and the Rate Principle in a different environment.

Example 2: A triathlete trains by running for two hours and biking for three hours on Tuesday, covering a total of 110 miles. The next day he runs one hour and bicycles four hours, covering 130 miles. How fast does he run? How fast does he bike?

Solution: Notice that we must assume he always runs at the same rate, which we shall denote R, and always bikes at the same rate, which we shall denote B. Using the Totals Principle, we see that the total distance traveled is the sum of the distance run and the distance biked. Using the distance = rate × time formula (the Rate Principle), on Tuesday he ran $2R$ miles and biked $3B$ miles, so $2R + 3B = 110$. On Wednesday, he ran R miles and biked $4B$ miles, so $R + 4B = 130$. The solution to these two equations is $R = 10$ miles per hour and $B = 30$ miles per hour. ■

The technique of elimination can be extended to more than two linear equations in more than two unknowns. Many reasonably priced pocket calculators can handle the solution of up to thirty equations in thirty unknowns, and computers can deal with thousands of equations in thousands of unknowns. However, as yet no computer, other than the human brain, can handle the much more important and difficult task of translating a real-world problem into the language of algebra. Artificial intelligence experts are working on this, but it's probably many years in the future.

MATHEMATICS OF FINANCE IN "MESSAGE FROM A CORPSE"

If you are a typical American, during your lifetime you will be involved with a substantial number of financial transactions. As a result, your ability to understand the mathematics of borrowing and lending is probably worth a minimum of tens of thousands of dollars to you, and possibly even more.

It is astounding how a person who carefully evaluates which of two types of TV to buy, where the cost of the two TVs may differ by at most a couple of hundred dollars, immediately accepts the first loan shoved under his or her nose by anyone who agrees to finance his or her purchase of a car or a house. Just as there are "best buys" in TVs, there are "best buys" in credit, for when you shop for a loan, or decide what to do with your investment capital, you are making a purchase of money.

This chapter can save you substantial amounts of money during your lifetime—far more than you might imagine—just by enabling you to see what borrowing money costs you. Since this chapter deals extensively with calculations, you would do well to own a calculator. A simple one, costing only a few dollars, that just has the arithmetic functions (+, −, ×, and ÷) will suffice, but life will be easier if you have one with exponentiation (this key is usually denoted y^x or x^y). It will only cost a few dollars more, and there's almost certainly one on your computer under "Accessories."

SIMPLE AND COMPOUND INTEREST

Many jobs pay wages that are computed by multiplying the length of time worked by the pay rate. If you are paid $8 per hour and work for 10 hours, you are paid $80. *Simple interest* can be thought of as the wages earned by money, and it is computed exactly the same way, by multiplying the length of time the money is working by the *interest rate*.

Example 1: Suppose that Sue loans John $500 for three years at an interest rate of 6% per year. How much does the money that Sue loans earn? How much does John have to pay back? Assume that simple interest is being charged.

Solution: Because 6% of $500 is 0.06 × $500 = $30, each year John has the money costs $30. Since he borrows it for three years, the money "earns" 3 × $30 = $90 during that period. John will have to repay the $500 he borrowed, plus the $90 interest, for a total of $590. ■

In example 1, the amount John borrows ($500) is called the *principal*. The basic time unit used for the computation of the interest rate (one year) is called the *interest period*. The interest rate (6% per year) is always given as a percentage per interest period.

The general formulas governing computation of simple interest are straightforward.

Rules for Simple Interest

If one borrows a principal P at a rate r per interest period for a total of t interest periods, then the total amount of simple interest I is given by

$$I = Prt \qquad [6.1]$$

The total amount A that must be repaid is

$$A = P + I = P + Prt = P(1 + rt) \qquad [6.2]$$

Although the rate *r* is always expressed as a percentage per interest period, when computing with it, remember to convert the percentage to a number (by dividing by 100).

If one is borrowing money, the principal *P* is sometimes called the *present value* of the loan, and the total amount *A* to be repaid is called the *future value*.

Notice that in example 1 the interest is added onto the principal, and the entire amount is paid back when the *loan period* (three years in example 1) expires. *Add-on interest* is the amount of simple interest that is added to the principal.

Compound Interest

Look again at example 1. Sue loans John $500 for three years, and John has the use of the entire $500 for all three years. An important feature of simple interest is that the recipient has the use of all the money loaned for the full period of the loan.

/ (Compound interest continued from p. 56)

Now let's take a look at the money in the Alma Steadman Trust in the story. If she deposits the original $2 million at 6% compounded annually, it is the same as loaning the money to the bank. At the end of every year, the money that was deposited at the start of the year, plus the interest that money has earned, is returned by the bank to Alma, and she immediately redeposits the new amount for another year.

Year	Balance on Jan. 1	Interest	Balance on Dec. 31
2004	$2,000,000.00	$120,000.00	$2,120,000.00
2005	$2,120,000.00	$127,200.00	$2,247,200.00
2006	$2,247,200.00	$134,832.00	$2,382,032.00
2007	$2,382,032.00	$142,921.92	$2,524,953.92
2008	$2,524,953.92	$151,497.24	$2,676,451.16
2009	$2,676,451.16	$160,587.07	$2,837,038.23
2010	$2,837,038.23	$170,222.29	$3,007,260.52
2011	$3,007,260.52	$180,435.63	$3,187,696.15
2012	$3,187,696.15	$191,261.77	$3,378,957.92
2013	$3,378,957.92	$202,737.48	$3,581,695.40

Notice that, after one year, the balance on 12/31/04 is $2,000,000 × 1.06. After two years, the balance on 12/31/05 is $2,000,000 × 1.06^2. After three years, the balance on 12/31/06 is $2,000,000 × 1.06^3. Finally, after 10 years, the balance on 12/31/13 is $2,000,000 × 1.06^{10}. So $2,000,000 is the original principal, 1.06 is 1 plus the annual interest rate, and 10 is the number of years that the deposit has been earning interest. This shows that the future value A of a principal P deposit at an annual interest rate r for N years is given by the formula

$$A = P(1 + r)^N \qquad [6.3]$$

This equation is extremely important. It involves four quantities: the principal P, the future value A, the annual interest rate r, and the number of years N that the money has been earning interest. If any three of these quantities are known, it is possible to solve for the fourth quantity in terms of the three missing quantities. For example, the present value P as a function of A, r, and N is given by

$$P = A/(1 + r)^N \qquad [6.4]$$

It is often important to compute the present value of an amount that will be needed in the future.

Example 2: Jose's parents decide to give him a car when he graduates from college in four years. If they estimate the cost of the car as $10,000, and a bank is paying 6% compounded annually, how much must they deposit now to be able to buy the car when Jose graduates?

Solution: We must find the present value P of an amount whose future value A = $10,000, when money is compounded at an annual rate r = 0.06 for four years. So

$$P = \$10,000/1.06^4 = \$7,920.94$$

We can check that this is correct by seeing that $7,920.94 deposited for four years at 6% compounded annually yields a future value of $10,000. ■

OTHER COMPOUNDING PERIODS

When money is compounded annually, the interest is computed and added on at the end of the year, and the new total is used as the principal for the next year. Other frequently used compounding periods are semiannual compounding (twice a year), quarterly compounding (four times a year), monthly compounding (12 times a year), and daily compounding (360 times a year). The fact that a banking year is only 360 days is a reminder of how difficult computation was B.C. (before calculators) because it is much easier to work with semiannual, quarterly, and monthly compounding when the year is 360 days rather than 365.

We could simply use formula [6.3] to compute the future value of any principal if we are given the interest rate r per compounding period. Then N would be the number of compounding periods.

Example 3: Meredith deposits $3,000 in a bank that pays a quarterly compounding rate of 1.5%. What is the amount in her account at the end of three years?

Solution: Since there are four quarters in a year, there will be 12 quarters in three years. Using formula [6.3],

$$A = \$3,000 \times 1.015^{12} = \$3,586.85 \ \blacksquare$$

Most of the time, however, a loan does not specify the interest rate r per compounding period but rather the annual compounding rate and the number of times per year that the loan is compounded. In example 3, the bank would say that it paid 6% compounded quarterly. The quarterly compounding rate is determined by dividing the annual compounding rate of 6% by 4, the number of compounding periods in a year.

This situation leads to the following modification of formula [6.3]. If a principal P is borrowed for N years at an *annual* rate r that is compounded t times a year, then the future value A is given by

$$A = P(1 + r/t)^{Nt}$$

Example 4: Suppose that $8,000 is deposited in a high-yield corporate bond for five years at 8% annually. Compute the amount in the account if compounding is done (a) annually, (b) semiannually, (c) quarterly, (d) monthly, and (e) daily.

Solution:

(a) $A = \$8,000 \times 1.08^5 = \$11,754.62$
(b) $A = \$8,000 \times 1.04^{10} = \$11,841.95$
(c) $A = \$8,000 \times 1.02^{20} = \$11,887.58$
(d) $A = \$8,000 \times 1.0066667^{60} = \$11,918.79$
(e) $A = \$8,000 \times 1.0002222^{1800} = \$11,933.59$

While (a), (b), and possibly even (c) can be done without a calculator that can handle exponentials, (d) and (e) are simply too much work unless such a calculator is available. ■

Example 4 makes two things clear. The first is that the more frequently money is compounded at the same annual rate, the greater the future value. The second is that either financial calculators, or calculators with an exponentiation key, are indispensable for computing when the compounding period is monthly or daily. Before the invention of electronic computers, the word "computer" did not refer to a machine but to an individual who was employed to do these sorts of calculations every day! A "computer" was a job description, not something that could be bought at an electronic supply store.

AMORTIZATION AND BUYING ON INSTALLMENT

New cars cost in the tens of thousands of dollars, and houses cost in the hundreds of thousands of dollars—and most of us simply don't have that much money on hand. Fortunately, there are companies whose business is to loan money to be repaid over a period of time—typically four or five years for cars, fifteen to thirty years for houses. The typical way that this is done is for the purchaser to immediately pay a fraction of the purchase price, known as the *down payment*, and pay the rest off periodically (usually monthly, but sometimes biweekly) in *installments*. Each installment payment is for the same amount.

Before diving into the formulas, let's see how this works for the first few payments in a typical case. Suppose you have your heart set on a new car that costs $20,000. The dealer asks for 10% down, so you write a check for $2,000. That leaves $18,000 to be financed, and the company supplying that $18,000 does so at a rate of 6%. You are then informed that you will make monthly payments of $422.73, and the first payment will be made a month from now.

The arithmetic of how this works is pretty straightforward. The moment you take possession of the car, you owe $18,000 to the credit company. At 6% annually, that's 0.5% a month. So, after the first month, you have borrowed $18,000 and the interest that has accrued is 0.5% of $18,000, or $90. Consequently, you now actually owe the company $18,090. When you make your first payment of $422.73, the amount you owe, called the *balance*, is reduced to $18,090 – $422.73 = $17,667.27.

When you made that first payment of $422.73, $90 went to pay off the interest, and the remainder, $422.73 – $90 = $332.73 was used to reduce the balance. That $332.73 is usually called the *principal*.

Another month goes by. You still have to pay 0.5% monthly on the balance of $17,667.27, but that's only $88.34. So, when you make that second payment of $422.73, the principal is $422.73 – $88.34 = $334.39; this reduces the balance to $17,667.27 – $334.39 = $17,332.88. And so it goes; each month the interest is less, the principal is larger, and the 48th payment reduces the debt to zero. The company now sends you the pink slip, and the car is all yours.

This method of paying off a loan is called *amortization*. There are many spreadsheets available online; you can either use them on a website or download them. Typically, a spreadsheet for the above loan will look something like this; because there are 48 separate payments, only the first three and the last three are shown.

Initial Balance = $18,000 Installment Payment = $422.73

Payment No.	Interest	Principal	Balance
1	90.00	332.73	17,667.27
2	88.34	334.39	17,332.88
3	86.66	336.07	16,996.81
. . .	. . .	. . .	. . .
46	6.28	416.45	839.21
47	4.20	418.53	420.68
48	2.10	420.63	0.05

As you can see, you still owe the company a nickel, and it's not going to let you get away with that! They'll ask you to make a last payment of $422.78 rather than $422.73. After all, if they let a million borrowers walk away with a nickel each, that's $50,000.

HOW INSTALLMENT PAYMENTS ARE COMPUTED

The formula for finding the amount of an installment payment comes from a clever bit of algebra.

Suppose that k is a constant, N is a positive integer, and we want to compute the sum

$$S = k + k^2 + k^3 + \ldots + k^N \qquad [6.5]$$

Multiply both sides by k.

$$kS = k^2 + k^3 + \ldots + k^{N+1} \qquad [6.6]$$

Subtract equation [6.6] from equation [6.5].

$$S - kS = (k + k^2 + k^3 + \ldots + k^N) - (k^2 + k^3 + \ldots + k^{N+1})$$

There's a lot of cancellation on the right side; what remains is

$$(1 - k)S = (k - k^{N+1})$$

So

$$S = (k - k^{N+1})/(1 - k) \qquad [6.7]$$

The basic principle used in figuring out your installment payments is that the sum of the present values of all the installment payments has to equal the initial balance. Suppose that the interest rate per period is given by i (in the example of the auto loan above, $i = 0.5\% = 0.005$ when expressed as a decimal). Then by formula [6.4], if each installment payment is P, the present value of the nth payment, the one made after n payment periods, is $P/(1 + i)^n$. The sum of the present values of installment payments 1 through N, which must be the current balance B, is therefore

$$B = P/(1 + i) + P/(1 + i)^2 + \ldots + P/(1 + i)^N$$
$$= P(k + k^2 + \ldots + k^N), \text{ where } k = 1/(1 + i)$$

We'll spare you the tiresome details of substituting equation [6.7] into this, solving for P, and simplifying. The result is that

$$P = Bi/(1 - k^N), \text{ where } k = 1/(1 + i) \qquad [6.8]$$

If you haul out a calculator with an exponentiation key and let $B = 18,000$ and $i = 0.005$ in the above formula, you will see that $P = 422.73$.

Maybe formula [6.8] doesn't get the PR that Einstein's $E = mc^2$ does, but formula [6.8] has a lot more impact because almost everyone will buy something on installment at some time in his or her life.

APPENDIX 7

SET THEORY IN "ANIMAL PASSIONS"

Set theory is a branch of mathematics that is very different from arithmetic or algebra. The fundamental questions of arithmetic and algebra deal with computation—what rules govern it, how best to perform it, and where to use it.

Although there are computational aspects to set theory, the primary use of set theory is to provide a framework in which to phrase and study a variety of important mathematical topics. It is impossible to conceive of mathematics without arithmetic and algebra, but almost all the branches of mathematics that use set theory were in existence long before the invention of set theory and were surviving quite nicely without it.

Nonetheless, once set theory had been invented, it was immediately put to use in a variety of situations.

SETS AND INCLUSION

The first order of business in constructing a new branch of mathematics is to define the objects we shall consider. A *set* is defined to be a collection of things. Of course, this raises the obvious question: what is a "thing"? Well, we can't define it, but we know what "things" are—a triangle, Abraham Lincoln, and the number 7 are all examples of things.

A set S can be specified by listing its things. For example, S = {1, 2, 3, 4} is the set consisting of the whole numbers 1, 2, 3, and 4. This method of listing the *elements* ("element" is a more impressive-sounding synonym for "thing") of a set works well if the set has only a few elements, but for sets with many elements, it is inefficient and is replaced by the "set-builder" method. To describe the set of all whole numbers between 1 and 1,000, one uses *set-builder nota-tion*. When set-builder notation is used, the "entrance requirements" for membership in the set are described by using letters as variables.

An example of describing a set using set-builder notation is to write the set S of all whole numbers between 1 and 1,000 as S = {$x : x$ is a whole number between 1 and 1,000}. This is read, "S is the set of all x such that x is a whole number between 1 and 1,000." The symbol x is sometimes called a *dummy variable* be-cause the particular symbol x plays no part in the actual set. We could have used any other symbol to denote the same set, such as

$$S = \{y : y \text{ is a whole number between 1 and 1,000}\}$$

or

$$S = \{☺ : ☺ \text{ is a whole number between 1 and 1,000}\}$$

Given a specific thing, which we shall denote by t, and a set S, we can ask whether or not t belongs to the set S. We write $t \in S$ to indicate that thing t belongs to set S, and $t \notin S$ to indicate that thing t does not belong to S. Note that drawing a diagonal line through the symbol $\in$ negates it. This is a standard mathematical convention, which we are already familiar with from arithmetic: $2 + 2 \neq 3$ means that 2 plus 2 does not equal 3. The same convention is followed with public information signs: A picture of a cigarette with a diagonal line through it means "no smoking." It is custom-ary to use capital letters, such as A, B, and C, to denote sets and lowercase letters, such as a, b, and c, to denote elements of sets.

Example 1: Formulate the sentence, "George Washington was a President of the United States who did not play baseball," using sets. Use set-builder notation, and $\in$ and $\notin$ symbols.

Solution:

Let A = {x : x was a President of the United States}
Let B = {x : x played baseball}
George Washington ∈ A, George Washington ∉ B. ■

Once a type of mathematical object has been defined (sets in our case), there are many questions that can be asked. Can we compare two such objects? How can we combine them? Of course, you are familiar with these questions for numbers, but these questions are among those typically asked when new mathematical objects are first studied.

Two sets A and B are equal if they contain the same elements. If A = {1, 2, 3, 4} and B = {3, 1, 4, 2}, these sets are equal (which we write A = B) because they contain the same things, even though they were listed in different orders. In this sense, a set is like a lunchbox because only the contents of the lunchbox matter—the order in which they were placed in the lunchbox, or the order in which they were removed from the lunchbox, is unimportant.

Subsets

We say that A is a *subset* of B if every element of A is an element of B. If A = {1, 2, 3} and B = {1, 2, 3, 4}, then A is a subset of B, written A ⊆ B. If A is a subset of B, we sometimes (but not often) say that B is a *superset* of A, written B ⊇ A. If A is not a subset of B, there must be some element of A that is not a member of B. Informally, if A ⊆ B, we say that A is contained in B, and B contains A.

Example 2: Let A = {p, q, r}. Is p a subset of A? Is {p} a subset of A? Is {r, q, p} a subset of A?

Solution: No, p is not a subset of A; it is an element belonging to A. {p} is the subset of A consisting of the single element p. Every element of {r, q, p} belongs to A, so it is a subset of A (it is, in fact, A itself, so A is a subset of A). ■

The subset relation between sets is in some ways similar to the "less than or equal to" (≤) relation between numbers. Both include equality as a possibility: We know that $8 \leq 8$, and also $A \subseteq A$ (every element of the set A on the left side of the $\subseteq$ symbol is certainly an element of the same set A on the right side of the $\subseteq$ symbol). Both symbols, $\leq$ and $\subseteq$, have horizontal bars under them to allow for the possibility that the object on the left of the symbol may be equal to the object on the right. This parallel among symbols can be extended to eliminate the possibility of equality by removing the horizontal bar: the symbol $3 < 5$ means that 3 is less than 5, and $A \subset B$ means that A is a subset of B but is not equal to B (sometimes we say that A is a *proper subset* of B).

Similarities between $\leq$ and $\subseteq$

Reflexivity: $a \leq a$ and $A \subseteq A$
Antisymmetry: If $a \leq b$ and $b \leq a$, then $a = b$
    If $A \subseteq B$ and $B \subseteq A$, then $A = B$
Transitivity: If $a \leq b$ and $b \leq c$, then $a \leq c$
    If $A \subseteq B$ and $B \subseteq C$, then $A \subseteq C$

The first property, reflexivity, is more an observation than something that is generally useful, but the last two properties are quite useful. The second property, antisymmetry, is often used to show that two sets are equal by showing that each is a subset of the other.

Universal Sets and Complements

Set theory provides a useful framework for discussing certain types of problems. However, often the items under discussion during these problems are limited: We might be discussing positive numbers, or presidents of the United States, or poker hands. In such situations, it is useful to have a universal set available, which may be loosely regarded as the "smallest" set containing all the items under discussion. If we are discussing positive numbers, for instance, we don't want to worry about presidents or poker hands, and so it would be sensible to define U as a universal set consisting of all possible positive numbers. It is then understood that all sets mentioned will be subsets of this universal set.

Unlike most of the concepts in mathematics, the concept of universal set is a little nebulous. For instance, if A = {Iowa, Illinois}, and B = {Illinois, Indiana}, possible universal sets are states or places beginning with the letter I.

Once a universal set has been defined, it makes sense to talk about all the objects in the universal set that are not in a given set. The *complement* of the set A, written A', is $\{x : x \in U, x \notin A\}$.

Example 3: Let U be the universal set consisting of positive whole numbers. If A = {1, 2, 3}, then describe A' in two different ways, using both words and set-builder notation.

Solution: A' is the set of all positive whole numbers greater than 3. Using set-builder notation, this would be A' = $\{x : x$ is a positive whole number greater than $3\}$. ∎

The complement of a set is a relative concept, in that it needs a universal set in order to make sense. In example 3, for instance, it is almost certain that we don't want to have to worry about hippopotamuses when considering the set A'!

The Connection between Logic and Set Theory

Set theory has an extremely useful connection with logic. If $P(x)$ is a proposition about the variable x (such as "x is a number greater than 7"), then P = $\{x : P(x)$ is true$\}$ is the set of all things for which $P(x)$ is a true statement. In this case, P would be the set of all numbers greater than 7. Similarly, we define Q = $\{x : Q(x)$ is true$\}$.

If both $P(x)$ and $Q(x)$ are propositions and the sets P and Q are defined as above, then the logical statement IF $P(x)$ THEN $Q(x)$ for all x supplies the same information as the set theory statement P $\subseteq$ Q.

For example, if $P(x)$ is the proposition that "x is a number greater than 7," and $Q(x)$ is the proposition that "x is a number greater than 4," clearly $P(x) \Rightarrow Q(x)$. Notice that if P = $\{x : x > 7\}$ and Q = $\{x : x > 4\}$, then P $\subseteq$ Q.

The Empty Set

The *empty set* is the set that contains no elements. If sets are lunchboxes, there's nothing in this lunchbox. The empty set is written $\emptyset$.

One of the most common mistakes is to confuse the empty set, $\varnothing$, with the number 0. As we shall see, there are certain similarities between the two objects, but they belong to different mathematical systems. It is as if one were to confuse the contents of an empty lunchbox (the empty set) with the cost of purchasing those nonexistent contents (zero).

The empty set has an extremely important property: it is a subset of *every* set! While this appears surprising at first, its proof simply involves the logical negation of the statement: $\varnothing \subseteq A$. If it is false that $\varnothing \subseteq A$, then there must be some element in $\varnothing$ that is not an element of A. But this is a contradiction because there are *no* elements in $\varnothing$—and once an assumption leads to a contradiction, that assumption must be false.

For example, the empty set is a subset of G, the set of all giraffes, for if it were not, there would be a giraffe in the empty set.

UNIONS AND INTERSECTIONS

We turn now to the question of how to combine two sets. Any interesting combination of two sets must itself be a set, just as the interesting ways to combine numbers (addition, subtraction, multiplication, and division) result in numbers. The two basic ways to combine sets are intersection and union.

The *intersection* of two sets A and B, written $A \cap B$, is the set of all elements common to both sets. If we define this using set-builder notation, $A \cap B = \{x : x \in A \text{ and } x \in B\}$.

Example 4: Let $A = \{1, 2, 3\}$ and $B = \{2, 3, 4\}$. Then $A \cap B = \{2, 3\}$.

If A and B have no elements in common, then A and B are said to be *disjoint*. In this case, $A \cap B = \varnothing$.

Example 5: If $A = \{1, 2, 3\}$ and $B = \{4, 5\}$, then A and B are disjoint, and $A \cap B = \varnothing$.

The other basic way of combining the two sets A and B, the *union* of A and B, written $A \cup B$, is the set of all elements that belong to either (or both) of the two sets. If we define this union using set-builder notation, $A \cup B = \{x : x \in A \text{ or } x \in B\}$.

Notice that the "or" used in defining the union of two sets is the inclusive "or," which is also the "or" of choice in logic. In day-to-day conversation, it is usually clear from context whether the inclusive or exclusive "or" is intended, but all the definitions of mathematics use the inclusive "or."

Example 6: Let A = {rat, cat, cow} and B = {dog, rat, cat, pig, yak}. Then A ∩ B = {rat, cat}, and A ∪ B = {rat, cat, cow, dog, pig, yak}.

In example 6, although "cat" appeared in both A and B, it is not listed twice in A ∪ B. When one is asked the contents of a lunchbox, it is repetitive to say, "a ham sandwich, uh, a ham sandwich, a bag of potato chips, and an apple."

When more than two sets are involved, parentheses are used to indicate which operations should be performed first, just as they are in arithmetic. Suppose that A = {Alan, Betty, Carol}, B ={Betty, David, Frank}, and C = {Alan, Carol, Edward}. Then (A ∪ B) ∩ C is determined first by computing A ∪ B = {Alan, Betty, Carol, David, Frank}, and then (A ∪ B) ∩ C = {Alan, Carol}. Notice that, if we were to compute A ∪ (B ∩ C), we would first determine B ∩ C = ∅, and then A ∪ (B ∩ C) = {Alan, Betty, Carol}. Therefore, the expression A ∪ B ∩ C must be parenthesized in order to be unambiguously determined.

Unlike arithmetic, where multiplication takes precedence over addition (so that, in the expression 2 + 3 × 4, the multiplication is performed first), there is no precedence hierarchy for the operations of union and intersection. Union and intersection are both *associative* operations, that is,

$$(A \cup B) \cup C = A \cup (B \cup C)$$
$$(A \cap B) \cap C = A \cap (B \cap C)$$

If only unions are involved, or intersections, it is not necessary to parenthesize, but if both operations appear in the same expression, you must parenthesize them in order to avoid ambiguity.

One possible reason that the empty set gets confused with the number 0 is that the empty set behaves in a similar fashion, relative to the operations of union and intersection, that the number 0 behaves relative to the operations of addition and multiplication.

$$(1) \ A \cup \varnothing = A$$

and

$$(2) \ A \cap \varnothing = \varnothing$$

If one regards the empty set as the set-theory analog of the number 0, then (1) is analogous to the arithmetic property $a + 0 = a$, and (2) to the arithmetic property $a \times 0 = 0$.

We have encountered both unions and intersections in the preceding story. If we let L denote the set of all members of Lisa's animal rights group who planned on attending the lobbying effort and R denote the set of members who planned on raiding the laboratory, then L ∪ R is the set of all the activists, those members who planned to participate in at least one of the two activities. L ∩ R consists of the "hard core" who planned to participate in both.

Many of the most interesting sets have only a finite number of things in them. If A is such a set, then $N(A)$ denotes the number of things in A. If A = {Alice, Betty, Carlos}, then $N(A) = 3$. Although there are many sets A for which $N(A) = 3$, there is only one set for which $N(A) = 0$, and that is the empty set.

The Fundamental Counting Principle

*(Fundamental Counting Principle continued from p. 64) In the preceding story, Pete makes use of an extremely important principle of counting.

If A and B are finite sets, then $N(A \cup B) = N(A) + N(B) - N(A \cap B)$

This valuable formula plays an important role in the study of probability, and so it is worth spending a little effort to verify it.

Pete's explanation was that, when we add $N(A)$ to $N(B)$, obtaining $N(A) + N(B)$, we are counting everything that appears in A ∩ B twice. Therefore, if we wish to compare $N(A \cup B)$ with $N(A) + N(B)$, we must realize that $N(A) + N(B)$ counts everything in A ∩ B twice. So $N(A) + N(B) = N(A \cup B) + N(A \cap B)$ since elements that belong to A or B, but not both, are counted once on each side of the equation, and elements that belong to A ∩ B are counted twice on each side of the equation. Subtracting $N(A \cap B)$ from both sides now yields the Fundamental Counting Principle.

Example 7: A survey of the 140 customers at an electronics supply store who owned either a high-definition TV or a GoPro camera revealed that 105 of them owned an HDTV and 28 owned both. How many owned a GoPro camera?

Solution: It is certainly possible to simply "plug into" the Fundamental Counting Principle. Let H be the set of HDTV owners, C the set of GoPro camera owners. Then $N(H \cup C) = 140$. Since $N(H) = 105$ and $N(H \cap C) = 28$, we have $N(H \cup C) = N(H) + N(C) - N(H \cap C)$, so $140 = 105 + N(C) - 28 = 77 - N(C)$. Therefore, $N(C) = 140 - 77 = 63$. ∎

The Difference of Two Sets

Another set that is of interest is the *difference* A\B, which is defined as the set of all things in A but not in B. That is, A\B = A ∩ B′, or, using set-builder notation, A\B = $\{x : x \in A \text{ and } x \notin B\}$. In examples 5 and 6, we used the idea of A\B before we had specifically defined it.

Example 8: If A = {dog, pig, rat, cat, yak} and B = {rat, cat}, what is A\B? What is B\A?

Solution: A\B is the set of all things belonging to A but not to B, which is {dog, pig, yak}. B\A is the set of all things belonging to B but not to A, which is ∅. ∎

We have mentioned the analogy between arithmetic and set theory. There is a temptation to regard the set-theoretic difference as being analogous to subtraction because when we construct the set A\B we are "taking away" the things in B from A. However, this analogy should not be taken too far. For example, if the number b is larger than the number a, the difference $a - b$ is negative. If the set B is "larger" than the set A (i.e., B ⊇ A), the set-theoretic difference A\B = ∅, the empty set. This situation occurs in the second part of example 7. There is no concept in set theory analogous to negative numbers.

THE CHINESE RESTAURANT PRINCIPLE: COMBINATORICS IN "NOTHING TO CROW ABOUT"

The need for counting techniques arises because it is often necessary to count very large numbers of items. After all, if there are only five or so items in a set, one just counts them. On the other hand, if there are several hundred items in a set, counting them is likely to take some time, and there is the possibility of making a mistake. It may even happen that the number of items we wish to count is so large that we would never be able to do it directly.

We have already discussed one of the most important counting procedures, the Fundamental Counting Principle, which is summarized in the formula

$$N(A \cup B) = N(A) + N(B) - N(A \cap B)$$

In this appendix, you'll learn about counting techniques based on *successive choices*. Many decisions can be viewed as a procedure based on successive choices. When we order dinner in a restaurant, we successively choose an appetizer, an entree, and a dessert. When a basketball coach selects a starting line-up, he or she chooses the center, the two forwards, and the two guards.

Mathematics started with the problem of counting, and counting problems still underlie many of the most important areas of mathematics, such as probability, statistics, and decision theory. The techniques in this section reappear throughout the remainder of this book.

WHERE THE CHINESE RESTAURANT
PRINCIPLE COMES FROM

The Chinese Restaurant Principle Pete referred to in chapter 8 derives its name from the following situation. A staple of Chinese restaurants has been the fixed-price meal, consisting of one appetizer chosen from a list of many different appetizers, and one main course from a similar list. It is traditional for the menu to present these choices in the form of two columns.

Menu

Column A	Column B
Won Ton Soup	Sweet and Sour Pork
Spareribs	Lemon Chicken
Rumaki	Oyster Beef
Egg Roll	Moo Goo Gai Pan
Dumplings	Shrimp with Lobster Sauce
	Mixed Fried Noodles
	Pressed Duck

For a fixed price, the diner gets to choose one dish from Column A and one from Column B. An obvious question now arises: how many different possible fixed-price meals can one have?

Two meals are different if they have either different appetizers, or different main courses, or both. There is a straightforward way to write out all the different possible meals. First write out all the different meals with won ton soup as the appetizer, then all the different meals with spareribs as the appetizer, etc. This appears below, abbreviating with dots (. . .) to simplify the task.

Won Ton Soup—Sweet and Sour Pork

. . .

Won Ton Soup—Pressed Duck

This results in seven different meals with won ton soup as the appetizer.

Spareribs—Sweet and Sour Pork

. . .

Spareribs—Pressed Duck

Likewise, seven more meals, all different from the ones with won ton soup as the appetizer. Similarly, there would be seven different meals with rumaki as the appetizer, seven different meals with egg roll as the appetizer, and seven different meals with dumplings as the appetizer. This makes a total of $7 + 7 + 7 + 7 + 7 = 5 \times 7 = 35$ different meals.

The number 35 above is obtained by multiplying the number of different possible appetizers (5) by the number of different possible main courses (7). Although the different meals were counted by listing them by appetizer, it would have made no difference to the final tally if we listed them by entrees. Sweet and sour pork was the main course in five different meals, as was lemon chicken, etc. Counting this way gives $5 + 5 + 5 + 5 + 5 + 5 + 5 = 7 \times 5 = 35$ different meals, just as before.

It is important to realize that the choice of main course is *independent* of the choice of appetizer. That is, once you have chosen an appetizer, you are perfectly free to choose any of the possible main courses (or vice versa, if you decide to choose the main course first). If the choice of one depends on the choice of the other (for instance, if the restaurant has a rule preventing you from ordering a poultry main course when you have chosen rumaki as an appetizer, or if you are not allowed to order two pork dishes, such as spareribs and sweet and sour pork), the choices are no longer independent, and the counting procedure used here is no longer valid.

The Chinese Restaurant Principle (even professional mathematicians use that phrase, possibly because they often congregate in such establishments because, let's face it, mathematicians are not rich and Chinese restaurants are generally not expensive) states that, if two choices can be made independently, the number of ways of making both choices is the product of the number of ways of making each choice separately. In the above example, "making both choices" corresponds to selecting a meal. The number of ways of making the appetizer choice is five, the number of ways of making the main course choice is seven, and so the number of ways of making both choices is $7 \times 5 = 5 \times 7 = 35$.

Two Choices

If two choices can be made independently, and there are p ways of making the first choice and q ways of making the second choice, than the number of ways of making both choices is pq.

Several Choices

Since no meal is complete without dessert, let's see what happens to the Chinese Restaurant Principle when you decide to have dessert as well. Suppose that the menu we were using before offers a complete dinner, including a choice of one appetizer from Column A (five choices), one main course from Column B (seven choices), and one dessert from Column C (three choices). How many different complete dinners are possible?

No lengthy analysis is needed to solve this one. A complete dinner can be viewed as the result of two independent choices: The first choice is the appetizer–main course combination, which you already know can be made in 35 different ways, and the second choice is the dessert, which can be made in three different ways. The Chinese Restaurant Principle thus enables us to conclude that there are $35 \times 3 = 105$ different complete dinners. Of course, $35 \times 3 = 5 \times 7 \times 3$.

As a result, the Chinese Restaurant Principle can be extended to any number of independent choices. If there are several different independent choices to be made, the number of different ways of making all choices is equal to the product of the number of ways of making each choice separately. This result is stated formally in the next section.

p Choices

Suppose that there are p independent choices to make, and the first choice can be made in N_1 different ways, the second in N_2 different ways, . . . , and the pth choice in N_p different ways. Then the number of different ways of making all choices is

$$N_1 \times N_2 \times \ldots \times N_p$$

It is easy to envision the Chinese Restaurant Principle as simply being the number of different p-course meals, where there are N_1 different first courses, N_2 different second courses, . . . , and N_p different last courses.

When the choices are not independent, you can often compute the total number of available choices by computing the number of possible choices and subtracting the number of excluded choices.

Example 1: Mammoth Tours offers fixed vacation packages from Los Angeles to either Honolulu, Acapulco, the Bahamas, or Puerto Rico. One can travel by boat or plane and can stay at a first-class, deluxe, or economy class hotel. How many different vacation packages are offered, if Mammoth Tours require that if you travel by boat, you must stay at a first-class hotel?

Solution: Since the choices of destination, method of travel, and accommodations are independent, by the Chinese Restaurant Principle there are $4 \times 2 \times 3 = 24$ different vacation packages. However, the number of excluded choices can also be computed by the Chinese Restaurant Principle. There are four excluded destinations and two excluded types of accommodation once you decide to travel by boat, so the number of excluded packages is $4 \times 2 = 8$. Therefore, the total number of packages offered is $24 - 8 = 16$. Alternatively, you could count the number of boat packages and plane packages separately and add them up. ■

APPENDIX 9

PROBABILITY AND EXPECTATION IN "THE WINNING STREAK"

Probability is a mathematical tool that has two primary functions. The first, and most straightforward, is to summarize existing information. The second, which is by far the more interesting, is to make predictions about the future. The crystal ball presented by probability is a little clouded, however, for the future that it uncovers is not a future of individual events but a future of long-term averages. The correct use of these long-term averages allows you to plan more sensibly for the future.

You undoubtedly have at least a passing acquaintance with probability through the weather reports. You've heard the TV meteorologist make a prediction such as this: The probability of rain tomorrow is 80%. Intuitively, you know that this means it's pretty likely that it will rain tomorrow. If you were to analyze this more thoroughly, you would conclude something like this: If we were to keep a running total of what the weather is actually like on all the days when the meteorologist says there is an 80% probability that it will rain, we would expect that it would rain on 80% of those days, but it would not rain on 20% of them.

The meteorologist's prediction helps you to make decisions. When the probability of rain the next day is 80%, you'll strongly consider taking an umbrella along with you, and you are unlikely to schedule a picnic. Yes, the meteorologist could be wrong—the day could be nice and sunny, and you would regret looking ridiculous as you carry your umbrella, and you might also regret not having a picnic on such a beautiful day. On the other hand, it would be a lot worse if it rained and you didn't have an umbrella and got soaked to the skin, and of course everyone knows what a disaster rain is when you have a picnic. You have no *guarantee* of making the right decision, but that 80% number is pretty convincing, so you take the umbrella and cancel the picnic. Of course, if the meteorologist said that the probability of rain was 100%, you'd have no problem deciding because it is certain to rain. Similarly, if the meteorologist said that the probability of rain was 0%, rain would have no chance of occurring, so the umbrella goes back in the closet and the picnic is on.

Probability, as you see it in predictions of rain, is a number that represents the *relative frequency* with which rain occurs: from a low of 0%, when rain never occurs, to a high of 100%, when rain definitely occurs. The higher the number, the more likely the rain.

THEORETICAL AND EMPIRICAL PROBABILITIES; SAMPLE SPACES

Summer days in Los Angeles belong to three basic types: sunny, rainy, and cloudy. Assume that these three types of days are defined to be nonoverlapping; a day is sunny, rainy, or cloudy. To make sure that the types of days are nonoverlapping, we might define a day to be rainy if any rain falls, cloudy if no rain falls but a cloud appears in the sky, and sunny otherwise. The weather forecast for the next day might be a 20% probability of rain, a 30% probability of sunshine, and a 50% probability that it will be cloudy.

The mathematical terminology that is used in the above example is that observing the next day's weather constitutes an *experiment*. The three possible *outcomes* to the experiment are sunny, which we shall abbreviate as *S*, cloudy (*C*), and rainy (*R*). The three possible outcomes, which must be nonoverlapping and cover all possibilities, constitute the *sample space* of the experiment.

Notice that the probabilities assigned to the three outcomes add to 100%: 20% + 30% + 50% = 100%. While it is perfectly possible to describe probability theory using percentages, there are technical reasons, which you'll see in a later section, that make it preferable to use numbers between 0 and 1, which we can obtain by dividing the percentages by 100. The probability of a rainy day becomes 0.2 after this division; this is usually abbreviated $P(R) = 0.2$. Similarly, we see that $P(S) = 0.3$ and $P(C) = 0.5$, and we observe that $P(R) + P(S) + P(C) = 0.2 + 0.3 + 0.5 = 1$.

Here's a summary of this.

Experiments, Sample Spaces, and Probabilities

An *experiment* is the observation of something that actually happens or could happen. The *outcomes* of this experiment are the different observations that might be made. The outcomes must be nonoverlapping and must cover all possible observations.

The *sample space* of the experiment is the set of outcomes. A *probability function* P is an assignment of numbers between 0 and 1 to the different outcomes in such a way that the sum of the probabilities of all the outcomes is 1.

The probability assigned to an outcome represents the relative likelihood that the outcome will actually occur when the experiment takes place. If the probability of an outcome is 0, the outcome cannot occur, and if the probability of an outcome is 1, then the outcome is certain to occur.

Experiments and sample spaces belong to one of three basic types, which depend on how the probabilities are assigned to the outcomes. There are sample spaces in which the probabilities are assigned on a subjective basis. The probability of acceptance of a marriage proposal is such an example, although generally the person offering such a proposal usually feels that this probability is close to 1. Exceptions, of course, exist—celebrities continually receive such proposals from hopeful, though probably not optimistic, strangers.

Probabilities can also be assigned on an empirical basis. If a quarterback has thrown 100 passes, completed 60, had 5 interceptions, and thrown 35 incomplete passes, one can assign probabilities by using the empirically determined relative frequencies: P(complete) = 60/100 = 0.6, P(interception) = 5/100 = 0.05, P(incomplete) = 35/100 = 0.35. In this instance, the probabilities can be interpreted as averages: for each pass thrown, the quarterback has averaged 0.6 of a completion, 0.05 of an interception, and 0.35 of an incompletion.

The third, and most readily analyzable, assignment of probabilities is the theoretical model. An easy example is the flip of a coin, in which the outcomes are heads and tails, denoted H and T, respectively. The symbol $P(H)$ denotes the probability of a head occurring as a result of the flip.

Example 1: The Fair Coin: One of the simplest experiments is the flip of a fair coin. There are only two outcomes: heads (H) and tails (T). To compute $P(H)$ and $P(T)$, the assumption that the coin is fair means that heads and tails are equally probable, so $P(H) = P(T)$.

Since H and T are the only possible outcomes, we must have $P(H) + P(T) = 1$. Since $P(H) = P(T)$, we obtain $P(H) = P(T) = 1/2$. This probably (appropriate word here!) comes as no surprise. ■

Expressed in terms of percentages, the probability of heads is 50%, and the probability of tails is also 50%. This idea has worked its way into the language: equal chances are sometimes called fifty–fifty chances.

The fair coin is a specific example of a uniform probability space. In a uniform probability space, all the outcomes are assumed to be equally likely (one of the meanings of "uniform" is "the same"). There are many interesting sample spaces in which the outcomes are all equally likely, such as the roll of a fair die or the drawing of a single card from a deck of 52 cards. It's easy to see, similar to the fair coin, that in a uniform probability space with n outcomes, the probability of each outcome is $1/n$.

When a fair die, which has six sides, is rolled, the probability of rolling a 4 is 1/6 (as is the probability of rolling a 1, or a 2, etc.). When a card is drawn from a deck of fifty-two cards, the probability of drawing the ace of spades is 1/52, which is the same as the probability of drawing the king of spades or the deuce of clubs.

7 (Probabilities for two-child family continued from p. 83)

Even Pete can make a mistake! Once March sees a boy, since he has no way of knowing whether he is seeing the younger or older child, it is twice as likely that he is seeing a family with two boys as either the BG or the GB family. There are four equally likely families: BB (he sees the younger boy), BB (he sees the older boy), BG, and GB—and in half those families the child March does not see is a girl. So DiStefano was offering odds of 11 to 10 on a 50–50 bet—but got lucky!

In the story, Pete was interested in the probability that the man March asked had a sister. This is not just a single outcome, such as BG, because we do not know whether we are asking the firstborn child. What Pete was interested in was whether the actual outcome belonged to a predetermined set of outcomes, which in this instance consisted of both BG and GB.

An event in a sample space is a set of outcomes; in other words, it is a subset of the sample space. In order to compute the probability that the actual outcome of an experiment belongs to a particular event, one adds the probabilities of all the outcomes in the event, just as Pete added P(BG) + P(GB) = 1/3 + 1/3 = 2/3 to (erroneously) get a probability of 2/3 that the man's sibling was a girl. It is customary to use capital letters to denote events. If we use the event from the story we have been discussing (the sibling of the man March asked was a girl), we might let A = {BG, GB}. So Pete computed that P(A) = P(BG) + P(GB) = 2/3. Generalizing this example, the probability of an event E, written P(E), is the sum of the probabilities of all the outcomes in the event E.

The null set, ∅, also represents a collection of outcomes: no outcomes, to be precise. When we perform an experiment, it is understood that something will be observed, so the probability that no outcome will be observed is 0. The null set is sometimes called the *impossible event*, or the *null event*, and P(∅) = 0.

If S is the entire sample space, the sum of the probabilities of all the outcomes in S is equal to 1, and so P(S) = 1. The entire sample space is sometimes called the *certain event* because it is certain that, when an experiment is performed, one of the outcomes will be observed.

Expectation

Now that we know how to compute probabilities, how do we use them to make plans? Many situations arise in which payoffs are associated with the occurrence of an outcome or an event. Sometimes these payoffs are measured in terms of money, as in the bet between March and DiStefano. Sometimes they are measured in other units. One can, for instance, compute the probability of a worker being sick and measure the payoffs in terms of hours worked, or the probability of a shipment of items containing defective items and measure the payoffs in terms of defective items.

Let's start with a simple example. Suppose that you draw a card from a deck of 52, and a benevolent individual offers to pay you $5 for each time you select a face card, provided that you pay her or him $1 every time you don't. The game goes as follows: you draw a card, pay or get paid, put the card back in the deck, shuffle the deck, and draw again.

If you were to play this game 52 times and happened to draw each card in the deck once, your balance sheet would show a gain of $5 for each of the 12 face cards, and a loss of $1 for each of the 40 other cards. Your net would be $(12 \times \$5) - (40 \times \$1) = \$20$. Since you played the game 52 times, your average win per play would be $20/52, which is approximately $0.38. We say that your *expected gain per play* is $0.38.

Let's look at a different way of computing the same number, $0.38. If we let F denote the event "draw a face card" and N denote the event "draw a nonface card," we see that our expected gain per play, which we know is $[(12 \times \$5) - (40 \times \$1)]/52$, can also be written

$$(12 \times \$5 - 40 \times \$1)/52 = 12/52 \times \$5 + 40/52 \times (-\$1)$$
$$= P(F) \times V(F) + P(N) \times V(N)$$

where V(F) denotes the value of getting a face card ($5), and V(N) denotes the value of getting a nonface card (−$1).

This setup gives us the "recipe" for computing the average gain per play associated with any game. Break the game up into events such that each outcome in a selected event has the same payoff. In the above game, the two events (F and N) were determined by the fact that each outcome in F had a payoff of $5, and each outcome in N had a payoff of -$1. Multiply the probability of each of the events by its associated payoff and add the result. This average gain per play is called the *expectation* of the game and is usually denoted by E. A game that has an expected value of 0 is called a *fair* game.

Definition of Expectation

Suppose that an experiment has outcomes $x_1, \ldots, x_N$.

If $P(x)$ denotes the probability associated with outcome x, and $V(x)$ denotes the payoff (value) associated with outcome x, then the expectation E is defined by

$$E = P(x_1)\, V(x_1) + \ldots + P(x_N)\, V(x_N)$$

Payoffs are positive from the point of view of the person or persons for whom the expectation is being computed. If several different outcomes have the same payoff, it is natural to group these together as an event. The above formula would then apply, with x_1 denoting the first event, and so on.

Once again, this is what Pete did in the story when he explained March's expectation by assuming that March bet $10 three separate times, winning once and losing twice. Denote the two events as S (the sex of the sibling is the *same* as the person being asked), and O (the sex of the sibling is *opposite* to the person being asked). According to the rules of the game, if March bet $10, $V(S) = \$11$ and $V(O) = -\$10$. You saw earlier that $P(S) = 1/3$ and $P(O) = 2/3$. Therefore, the expectation is

$$E = P(S) \times V(S) + P(O) \times V(O) = 1/3 \times \$11 + 2/3 \times (-\$10) = -\$3$$

You may recall that Pete described the expectation as 30% (from DiStefano's point of view). It wasn't clear whether each bet was $10, $100, or $1,000, so one might just as well assume that each bet was 10 units. Then the expected loss per bet would be 3 units.

Considering that 10 units was the amount of the bet, the percentage expected loss per unit bet would be 30%. Percentages therefore give a convenient way to describe the expectation of a game.

Odds

When a game has only two outcomes, sometimes the payoffs are quoted in terms of how much the player will win as compared with how much the player will lose. Odds of 3 to 1 means that the player stands to win three units if he or she wins, but will lose one unit if he or she loses. Odds of 3 to 1 describes those situations in which the player stands to win three times as much as he or she stands to lose.

Odds of 2 to 5 means that the player stands to win two units but risks losing five units. It is customary to quote odds in whole numbers at the simplest possible ratio. Odds of 1 to 1 are sometimes called *even money*.

Example 2: A British bookmaker (bookmaking is not only legal in England but respectable) has had punters (British for bettors) bet 300 pounds on Oxford and 200 pounds on Cambridge on the upcoming boat race. If the bookmaker plans to take 50 pounds for his or her commission, what odds should he or she offer on each school?

Solution: A total of 500 pounds has been bet, and after he or she takes his or her commission, 450 pounds are left. Those who have bet on Cambridge stand to lose 300 pounds, but they could win 450 pounds. So the odds the bookmaker should offer for a bet on Cambridge are 450 to 300, or 3 to 2. Similarly, since 200 pounds were bet on Oxford, the odds on Oxford are 450 to 200, or 9 to 4. ∎

A reasonable assumption in example 2 is that because 0.6 of the total money wagered was bet on Oxford, the probability that Oxford will win the race is 0.6. In this case, the expectation of the Oxford bettors is

$$E = (0.6 \times 150) + (0.4 \times -300) = -30$$

In this case, though, the percentage expectation of the Oxford bettors is therefore

$$100 \times -30/300 = -10\%$$

But it is not just games that have expectations associated with them. When an insurance company sells a policy or a company brings out a new product, they compute expectations. To decide whether an investment is a good one, they must calculate the expectation involved. Sometimes the probabilities or payoffs must be estimated based on lack of complete knowledge. A company usually brings out a new product only if it can project a positive expectation.

APPENDIX 10

CONDITIONAL PROBABILITY IN "ONE LONG SEASON"

↑ (Conditional probability continued from p. 93)
Let's analyze Julie's dilemma (to stand pat or to switch her choice) from the story at the start of chapter 10. Recall that Julie has initially selected Wyatt as the man she thinks Debbie will marry. The scriptwriters have already decided whom Debbie will marry before Julie even got involved. There would be no drama, from the standpoint of Silktex shampoo, if Wyatt was in a coma up in Monterey—nobody would watch to see Julie's choice because Wyatt was now clearly out of the picture. So the scriptwriters were told to involve either Ellison or Lowell, neither of whom was Julie's choice, in the car crash.

Since we have no advance information about which one Debbie is likely to marry, we can assume that she is equally likely to marry Ellison, Wyatt, or Lowell.

> *Case 1*: Debbie marries Wyatt. Either Ellison or Lowell would be in the car crash. It is wrong for Julie to switch (she would lose the $5,000 cost to switch, as well as the grand prize of $100,000).

Case 2: Debbie marries Lowell. The scriptwriters were obviously told to involve Ellison in the car crash. Now Julie wins the $100,000 by switching.

Case 3: Debbie marries Ellison. In this case, the scriptwriters would have been told to make Lowell the victim of the car crash. Again, it is right for Julie to switch.

The key point here is that the scriptwriters are guaranteed to put either Ellison or Lowell in the car crash once Julie selects Wyatt as Debbie's intended. We can therefore see that, in two out of these three cases, it is right for Julie to pay the $5,000 and switch choices.

The scriptwriter was given the following *condition*: Involve Ellison or Lowell in a car crash. In so doing, the sample space of the experiment has been modified from the original sample space (Debbie might marry any of the three suitors) to a subset of the original sample space (Debbie will only marry either Wyatt or Lowell). Because Ellison was in a car crash, the original sample space {Wyatt, Ellison, Lowell} has been changed; it's now {Wyatt, Lowell}. The study of sample spaces that have been modified as a result of additional information forms the subject of conditional probability.

The argument that Julie should switch is far from obvious and indeed goes against the grain for most people. When a variation of this problem (known by mathematicians as the Monty Hall Problem) was posed in a nationally syndicated magazine column, the author of the column mentioned that it generated incredible amounts of write-in commentary, and some very highly educated people came to the wrong conclusion!

CONDITIONAL PROBABILITY

The poker game of five-card stud was the de facto money poker game for almost a century until televised big-stakes poker made Texas Hold 'Em the game of choice.

In Texas Hold 'Em, each player is initially dealt two cards face down so that only the player to whom those cards are dealt can see them. After a round of betting, three cards are simultaneously turned face up in the center of the table (this is called "the flop"), and each player regards those three cards as augmenting the two they were originally dealt.

In five-card stud, each player is initially dealt one card face down, which only the player can see, and one card face up, which everyone can see. A round of betting ensues. Then each player is dealt a card face up, and another round of betting ensues. This continues either until all but one player has "folded" (refused to match a bet made by another player) or until the remaining players have a total of five cards, one face down and four face up.

Suppose that in a game of five-card stud, you are dealt the ace of clubs face down, and the ace of hearts and the four, five, and six of diamonds face up—a pair of aces. Doc, your opponent, has been dealt the five, seven, nine, and queen of spades face up. You quickly observe that Doc can only win if he has a spade face down to give him a flush, and you also quickly calculate that there are 43 unseen cards, of which 9 are spades. Based on only this information, Doc's chances of winning are therefore 9/43, and yours are a healthy 1 − 9/43 = 34/43.

Doc has, however, made a serious mistake: he is sitting with his back to a large mirror, and as he glances at his down card you can see a flash in the mirror—not enough to know for certain what that card is, but enough to convey some information to you. Let's look at three different cases.

> *Case 1*: You catch a glimpse of black. Uh-oh. Doc's card must now be one of 21 unseen black cards (you have seen your ace of clubs and Doc's four spades of the 26 black cards in the deck). Since nine of them are spades, Doc's chances of winning have increased to 9/21 = 3/7, and yours have decreased to 1 − 3/7 = 4/7.
>
> *Case 2*: You see a flash of red. Doc's card must now be one of 22 unseen red cards (you have seen your ace of hearts and four, five, and six of diamonds), and no red cards can be spades! Doc's chances of winning are 0/22 = 0, so you are certain to win! Gleefully, you watch Doc finger his chips, hoping he'll try to bluff you out of the pot.
>
> *Case 3*: You see the markings indicating a face card. Doc's card must now be one of 11 unseen face cards (you have seen Doc's queen of spades), two of which are spade face

cards (the jack or the king). Doc's chances of winning are therefore 2/11, and yours are 1 – 2/11 = 9/11.

Each of these situations constitutes a problem in conditional probability. In the original situation, you had no information on the nature of Doc's card. The sample space S for the experiment was therefore *all* forty-three unseen cards in the deck, and the event D (Doc has a winning hand) consisted of all nine unseen spades. Since this is a uniform probability space, P(D) = N(D)/N(S) = 9/43.

In each of the above three cases, information came to you that changed the sample space for the experiment. The *new* sample space was a subset of the original sample space S. The event that Doc wins was also altered by the outcomes available in the new sample space. In case 3, for instance, the new sample space was F, the set of all unseen face cards, and the event that Doc wins consisted of all spades that were also unseen face cards. The eight of spades, which was a winning card for Doc in the original situation, was no longer a winning card in light of the fact that it could not belong to F, the revised sample space.

THE PROBABILITY OF A GIVEN B

Let's look at this from a more general standpoint. Suppose that there are two events A and B in a sample space S. We now define $P(A \mid B)$ to be the probability of the event A, given that the event B has already occurred (the symbol $A \mid B$ is read "A given B"). The conditional probability $P(A \mid B)$ is given by the following formula.

$$P(A \mid B) = P(A \cap B)/P(B) \qquad [10.1]$$

Notice that this definition gives the same results as the computations we have already done in the three cases discussed in the poker problem. For instance, in case 3, if we let D be the event that Doc wins (Doc's unseen card is a spade) and B be the event that Doc's unseen card is a face card, then $D \cap B$ is the event that Doc's unseen card is a spade face card. We had already computed above that $P(D \mid B) = 2/11$. Since N(B) = 11 and $N(D \cap B) = 2$, in the original sample space S (no information about the unseen card), we see that P(B) = 11/43 and $P(D \cap B) = 2/43$. Therefore, the computational rule $P(D \mid B) = P(D \cap B)/P(B) = (2/43)/(11/43) = 2/11$.

APPENDIX 11

STATISTICS IN "THE GREAT BASKETBALL FIX"

LIES, DAMNED LIES, AND STATISTICS

You're constantly exposed to statistics. On a typical day, you might receive statistical information from the stock market (the Dow Jones averages), the entertainment industry (the Nielsen ratings), sports (baseball batting averages), economic reports (cost of living indexes), and a multitude of other areas. Everybody uses statistics to make a point, generally the point they want to make.

As a result, statistics could probably use some good PR because many people feel that statistics are used to promote a particular point of view and cover up the truth. That viewpoint is expressed in the sentiment so brilliantly expressed by Disraeli: "There are three kinds of lies: lies, damned lies, and statistics."

Statistics are frequently used to summarize data so that the data become more useful. In a world increasingly devoted to collecting and processing data, you can be overwhelmed by the data tsunami. It is not possible to understand a mass of data in raw form, not because it is intrinsically incomprehensible, but simply because there is so much of it. To understand the results of the 2012 U.S. presidential election, nobody wants to know how all 100 million voters, as individuals, voted, or even the vote totals for Obama and

Romney. You want to know the percentage of voters who voted for each candidate.

These percentages constitute one of the fundamental tools of statistics: the *probability distribution*. Recall that probability had both an empirical and a predictive aspect, and the same can be said of statistics. The two basic problems of statistics are how to summarize data and how statistics can be used to make intelligent decisions.

RANDOM VARIABLES, DISTRIBUTIONS, AND GRAPHS

Throughout this section, S is the sample space of an experiment whose possible outcomes are $O_1, O_2, \ldots, O_n$. During a basketball game one evening, LeBron James made eight two-point field goal attempts and missed seven, and made two three-point field goal attempts and missed three. The sample space of this experiment is an attempted field goal by LeBron James, and the possible outcomes are

O_1 = made a 2-point field goal
O_2 = missed a 2-point field goal
O_3 = made a 3-point field goal
O_4 = missed a 3-point field goal

This information can be obtained from the box score in the paper the next day. It can be summarized in terms of a function X whose domain consists of the numbers 1, 2, 3, and 4 (the possible outcomes) and whose values are given by

$$X(1) = 8 \qquad X(2) = 7 \qquad X(3) = 2 \qquad X(4) = 3$$

The function X defined here, which is associated with the sample space S, is called a *random variable*. A random variable is simply a function whose domain (the allowable inputs) are outcomes from a sample space and whose range (the outputs of the function) are numbers. It is customary to use capital letters at the end of the alphabet, such as X, Y, and Z, to denote random variables.

When the range of a random variable consists of nonnegative whole numbers, the term *frequency distribution* is used to describe the random variable. The random variable X defined above is a frequency distribution. $X(1)$ is the frequency of two-point field goals LeBron James made, X(2) is the frequency of two-point field goals LeBron James missed, etc.

In the basketball game above, $X(1) + X(2) + X(3) + X(4) = 8 + 7 + 2 + 3 = 20$, which is the number of field goals LeBron James attempted. Define $Y(1) = X(1)/20 = 8/20 = 0.4$; then a probabilistic interpretation of $Y(1)$ is that, if you choose a LeBron James field goal attempt at random, that attempt would be a successful two-point field goal with probability 0.4. (The percentage interpretation would be that 40% of LeBron James's field goal attempts were successful two-pointers.) Similarly, if $Y(2) = X(2)/20 = 0.35$, $Y(3) = X(3)/20 = 0.1$, and $Y(4) = X(4)/20 = 0.15$, then $Y(1) + Y(2) + Y(3) + Y(4) = 0.4 + 0.35 + 0.1 + 0.15 = 1$. The function Y is simultaneously a random variable and a probability function. A random variable that is also a probability function is called a *probability distribution*.

Let X be the random variable associated with LeBron James's field goal attempts. By borrowing the letter P from probability, you can discuss the various probabilities associated with LeBron James's field goal attempts without having to introduce another letter (such as Y). The notation

$$P(X = 1) = 0.4$$

is read "the probability that the random variable X assumes the value 1 is 0.4." This notation is commonly used to describe probability distributions associated with frequency distributions.

Example 1: Rutabaga Biotech has twenty-seven employees whose salaries are less than $40,000 a year, sixteen employees whose salaries are between $40,000 and $80,000 a year, and seven employees whose salaries are more than $80,000 a year. Describe the sample space, frequency distribution, and probability distribution associated with this experiment.

Solution: O_1 = an employee has a salary of less than $40,000 a year, O_2 = an employee has a salary of between $40,000 and $80,000 a year, and O_3 = an employee has a salary of more than $80,000 a year. The frequency distribution is the random variable X such that

$$X(1) = 27 \qquad X(2) = 16 \qquad X(3) = 7$$

Since $27 + 16 + 7 = 50$, the probability distribution associated with this random variable is $P(X = 1) = 27/50 = 0.54$, $P(X = 2) = 16/50 = 0.32$, and $P(X = 3) = 7/50 = 0.14$. ■

In example 1, there's a choice of how to describe the outcomes of the sample space. One possibility was to describe the possible outcomes as O_1 = a salary of \$1/year, O_2 = a salary of \$2/year, etc., up through the maximum salary that an employee at Rutabaga Biotech makes. This has the obvious inconvenience of having an experiment with more than 80,000 different outcomes. Alternatively, only the actual salaries could have been used as possible outcomes, which would limit the sample space to 50 outcomes if everyone had a different salary. The actual procedure selected, using a range of possible values as a particular outcome, is known as *binning* (shorthand for "to place in a bin"). In example 1, a judicious choice of bins has made it relatively clear what the salary structure at Rutabaga Biotech is, using a sample space with only three different outcomes.

MEASURES OF CENTRAL TENDENCY AND DISPERSION

During the height of a political campaign, one is deluged with "sound bites," those little morsels that condense an extremely complex position into one memorable phrase or slogan. Because frequency distributions can also be complex, there is substantial interest in finding numbers that can serve as "data bites," compressing much of the information of the distribution into a very few quantities. There are two basic types of "data bites": measures of *central tendency*, which locate the middle of the distribution, and measures of *dispersion*, which tell how tightly packed the distribution is around its middle.

The Mean: A Measure of Central Tendency

The *mean* of a distribution is just our old friend, the average. If you are given numbers $X_1, \ldots, X_n$, the mean, denoted by μ (the Greek letter mu), is simply

$$\mu = (X_1 + \ldots + X_n)/n \qquad\qquad [11.1]$$

Example 2: In six consecutive home games, the Chicago Cubs pitching staff allowed 4, 8, 9, 6, 6, and 9 earned runs (the wind was blowing out in Wrigley Field). Calculate the mean of this distribution.

Solution: The mean μ = (4 + 8 + 9 + 6 + 6 + 9)/6 = 7. ∎

The Standard Deviation: A Measure of Dispersion

The standard deviation, which is denoted by σ (the Greek lower-case sigma), is defined according to the equation

$$\sigma = \sqrt{\frac{(X_1 - \mu)^2 + (X_2 - \mu)^2 + \ldots + (X_n - \mu)^2}{n - 1}} \qquad [11.2]$$

This rather imposing looking formula is calculated by the following procedure.

Step 1: Find the mean.
Step 2: Compute the sum of the squares of all the deviations.
Step 3: Divide the result of step 2 by one less than the number of data points.
Step 4: σ is the square root of the result of step 3.

This really isn't a problem any longer since computers can calculate standard deviations easily (it's built into most spreadsheets, such as Excel), and it is not much bother with a hand calculator. In fact, many handheld calculators are programmed to find the standard deviation of a data set. You simply key in the data points, and the result appears.

Example 3: What is the standard deviation of the number of runs in example 2?

Solution: The mean was already computed to be 7, and the deviations to be –3, –1 (twice), 1, and 2 (twice). The squares of the deviations are 9, 1 (three times), and 4 (twice). The sum of the squares of the deviations is 20. Since there are 6 data points in the sample, divide by 6

– 1 = 5, obtaining 4. Finally, take the square root of this number, obtaining σ = 2.00. And you didn't even need a calculator for this one!

Although the standard deviation appears unnaturally complicated, it has the great advantage that predictions can be made from it.

THE NORMAL DISTRIBUTION

When Pete sent Freddy to report on how many free throws Theresa Middlebury made out of 100, he knew that Freddy would report that Theresa would have made a whole number of free throws. She would not have made 72.1 or 86.437 free throws. The number of free throws Theresa made out of 100 is an example of a *discrete* random variable, a random variable that can only assume a fixed, finite number of values (in Theresa's case, between 0 and 100).

In contrast, had Pete sent Freddy to measure Theresa's height, any value would conceivably have been possible, such as 64.18 or 67.304 inches, assuming that Freddy had a sufficiently finely calibrated ruler. Admittedly, you normally measure height to within the nearest inch or half inch, but this is by choice—it is certainly within our capability to measure more accurately. A random variable that can assume any value (within a given range) is said to be *continuous*.

Many continuous random variables have a distribution shaped like the one in figure 11.1, which is a hypothetical distribution of the heights of basketball players. The curve is known as a *probability density function*. It has the property that the total area under the curve is 1. The probability that a randomly selected basketball player, from a distribution with a mean of 78 inches and a standard deviation of 4 inches, has a height between 74 and 81 inches is the total amount of area shaded in figure 11.1. This could also be interpreted as the fraction of basketball players in the distribution between 74 and 81.

The bell-shaped curve in figure 11.1 is a special type of distribution known as a *normal* distribution. Normal distributions are extremely important because not only are they characteristic of many continuous random variables, but they also provide excellent approximations to many discrete random variables (Pete talked about this in the story).

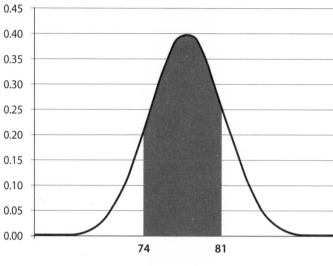

Figure 11.1

There can be many different distributions with mean μ and standard deviation σ. However, there is only *one* normal distribution with mean μ and standard deviation σ. Consequently, when you know the mean and standard deviation of a normal distribution, you know precisely which bell-shaped curve to draw, since there is only one. Once the curve is drawn, you can answer any question about the distribution.

BINOMIAL DISTRIBUTIONS

/ (Binomial distribution continued from p. 102)
In the story in chapter 11, you learned that Theresa Middlebury was an 80% foul shooter (reasonably good, even by NBA standards!). What is the probability that she would make precisely one out of two free throws?

When analyzing this problem, it is necessary to assume that each free throw is independent—no matter whether she hits or misses the first, her probability of sinking the second will still be the same: 0.8. When two events are independent, recall that the probability of *both* events occurring is obtained by multiplying the probability that *each* event will occur. The probability that she will make the first shot and miss the second is 0.8 × 0.2 = 0.16. Similarly, the probability that she will miss the first shot and make the second is 0.2 × 0.8 = 0.16. So the probability that she will make precisely one out of two is 2 × 0.16 = 0.32.

Now let's consider the probability of Theresa making exactly eighty out of a hundred free throws. One way she could do this is to make the first eighty and miss the last twenty. Since these free throws are all independent, the probability is obtained by multiplying $0.8 \times \ldots \times 0.8 \times 0.2 \times \ldots \times 0.2 = 0.8^{80} \times 0.2^{20}$. Another way that she could make exactly eighty out of a hundred free throws would be to miss the first twenty and then make the remaining eighty. The probability of this happening is $0.2 \times \ldots \times 0.2 \times 0.8 \times \ldots \times 0.8 = 0.2^{20} \times 0.8^{80} = 0.8^{80} \times 0.2^{20}$. A third way this could happen would be for her to miss the first ten, make eighty in a row, and then miss the last ten. The probability of this happening is $0.2 \times \ldots \times 0.2 \times 0.8 \times \ldots \times 0.8 \times 0.2 \times \ldots \times 0.2 = 0.2^{10} \times 0.8^{80} \times 0.2^{10} = 0.8^{80} \times 0.2^{20}$. No matter in which specific order you arrange for Theresa to make eighty and miss twenty, the probability that she will shoot the free throws in that exact order is $0.8^{80} \times 0.2^{20}$.

To compute the probability that Theresa will make precisely eighty out of a hundred free throws, you must therefore add $0.8^{80} \times 0.2^{20}$ once for each of the different orders in which she could make eighty and miss twenty. How many different such orders are there? If you imagine that you have a hundred numbered balls in a jar, and that you choose eighty of them, you could simply decree that Theresa makes the free throws whose numbers are on the balls that have been chosen and misses the others. Therefore, if you choose the eighty balls numbered 11 through 90, Theresa would miss the first ten, make shots 11 through 90, and miss the last ten.

The number of ways of choosing eighty numbered balls from a hundred is 100!/(80! × 20!) [factorials work this way: $n! = 1 \times 2 \times 3 \times \ldots \times n$]; the reasoning behind this calculation can be found at press.princeton.edu/titles/10559/html. Therefore, Theresa's probability of making exactly eighty of a hundred free throws is $(100!/(80! \times 20!)) \times 0.8^{80} \times 0.2^{20}$, which is approximately 0.0993.

There are really only three essential numbers in the above formula, as the other numbers can be computed from knowing these three. The first number is 100, which represents the number of free throws Theresa shoots. The next number is 80, which represents the number of free throws Theresa makes. The third number is 0.8, which represents the probability of Theresa making a free throw. The number 0.2, which is the probability of Theresa missing a free throw, is $1 - 0.8$, and the number 20, which is the number of free throws Theresa misses, is $100 - 80$.

Now let's generalize this a little. Suppose that you send a random free-throw shooter to the line N times. You assume that her probability of making each free throw is p and that each shot is independent of the others. The probability that she will make exactly k free throws (and miss exactly $N - k$) is given by the formula

$$C(N,k) \times p^k \times (1-p)^{N-k}, \text{ where } C(N,k) = N!/(k! \times (N-k)!) \quad [11.3]$$

This formula, which is known as the *binomial distribution formula*, can be used in a much wider context. The binomial distribution is the theoretical probability distribution in which

$$P(X = k) = C(N,k) \times p^k \times (1 - p)^{N-k} \quad [11.4]$$

Bernoulli Trials

Suppose you conduct an experiment that has only two outcomes, one of which you call *success*, which occurs with probability p, and (naturally enough), the other is called *failure*, which occurs with probability $1 - p$. Suppose further that each trial of this experiment is independent of the other trials; this situation is known as a Bernoulli Trials experiment. The probability of getting *exactly* k successes in N trials is given by formula [11.4].

$$P(X = k) = C(N,k) \times p^k \times (1 - p)^{N-k}$$

Earlier you obtained the probability of Theresa making exactly eighty of a hundred free throws as $C(100,80) \times 0.8^{80} \times 0.2^{20} = 0.0993$. There is a tremendous amount of calculation involved in obtaining the number 0.0993. (WARNING! Even with a pocket calculator, this calculation is time-consuming. Additionally, if the numbers are not computed in the correct order, it will take you beyond the limit of your calculator; 100! is often beyond the limit of a typical pocket calculator.)

Example 4: Assume that Theresa is an 80% foul shooter. What is the probability that she will make at least eighty out of a hundred free throws?

Solution: Using the same reasoning as in example 1, it is easy to see that this probability is the sum of the probabilities that she will make exactly eighty out of a hundred, and the probability that she will make exactly 81 out of a hundred, . . . , and the probability that she will make exactly a hundred out of a hundred (the NBA record for consecutive free throws, as of 2012, is 97). This number is

$$C(100,80) \times 0.8^{80} \times 0.2^{20} + \ldots + C(100,100) \times 0.8^{100} \times 0.2^{0}$$

Computing the value of this number is such a strain that we will temporarily abandon the project until we find a shortcut. ■

Fortunately, a shortcut looms just over the horizon.

The Normal Approximation to a Binomial Distribution

Mathematicians are no different from anyone else; they like to do as little grunt work as possible. It was discovered more than a century ago that the normal curve provided a good approximation to the binomial distribution as long as both Np and $N(1 - p)$ are both greater than 5. The normal curve that best matches the binomial distribution has mean $\mu = Np$ and standard deviation $\sigma = Np(1 - p)$.

Recall that this is exactly what Pete did in the story. Freddy reported back that Theresa had made sixty-six out of a hundred free throws. Freddy's attitude was, "Oh well, Theresa's just having an off day." Pete, however, approximated the binomial distribution with a normal distribution whose mean $\mu = 100 \times 0.8 = 80$ and whose standard deviation $\sigma = 100 \times 0.8 \times 0.2 = 16 = 4$. This was valid because $Np = 100 \times 0.8 = 80 > 5$, and $N(1 - p) = 100 \times 0.2 = 20$, which is greater than 5. Since 66 was 14 below the mean of 80, and 14/4 = 3.5 standard deviations, Pete realized that the probability of Theresa's having that bad an off day was less than one chance in a thousand, and so something other than an "off day" had to be going on.

To finish off example 4, because 80 is the mean of the distribution, the probability that Theresa will make at least eighty free throws is computed by looking at the normal distribution and seeing what percentage of the distribution lies above 79.5 (in approximating free throws made, which is a discrete variable, by a continuous variable, one thinks of "making precisely eighty free throws" as making more than 79.5 free throws but less than 80.5 free throws). The probability that she will make at least eighty free throws, using this method, is actually about 55% because 79.5 = $\mu - 0.125\ \sigma$.

GAME THEORY IN "IT'S ALL IN THE GAME"

In the story, Freddy found himself faced with a choice of two possible actions, or *strategies*: He could either call Lisa, or he could decide not to call her. How his choice worked out did not depend solely on what strategy he chose, but also on how Lisa felt. Lisa either wanted Freddy to call or she didn't, but it is probably incorrect to say that she is an opponent bent upon making life as miserable for Freddy as possible. When the opposing actions are not the result of a conscious choice, they are usually referred to as *states*, rather than strategies. The *payoff* Freddy received (which is measured in what might be called relative happiness points, with a high of 10 and a low of 0) depended on the strategy he chose and the state Lisa happened to be in. Since each of the two principals has two possible choices, this is called a *2 × 2 game*. If Freddy expanded his list of strategies to include a surprise visit to New York, the result would be termed a *3 × 2 game*. It is customary to use the first of the two numbers to denote the number of strategies available to the player from whose standpoint we measure the payoffs.

Game theory was started in the 1920s by the mathematician Emile Borel, although the first significant work was a book published in 1944 called *Theory of Games and Economic Behavior*, by John von Neumann and Oskar Morgenstern. Because many conflict situations arising in war can be conveniently formulated in the framework of game theory, it was intensively investigated during World War II. However, game theory has found a large number of applications to areas probably not considered by its founders. This is characteristic of mathematics—applications pop up in surprising places.

2 × 2 GAMES

(2 × 2 games continued from p. 112)

Saddle Points

Let's take another look at the diagram that Pete made for Dolores.

	High Price (thousands)	Reserve Price (thousands)
Classic	100	20
Vintage	70	30

If Dolores auctions the card at Classic, the worst that can happen is that she receives 20 points (recall that each point is equal to $1,000). This number is called a *row minimum* (it is the minimum of all the numbers in the first row). Similarly, if she auctions the card at Vintage, the worst that can happen is that she receives thirty points. The "best of the worsts," the larger of the two row minimums, is thirty points. This number is called the *maxmin*, which is an abbreviation for the maximum of the row minimums. The maxmin of thirty points occurs when she auctions the card at Vintage.

Now look at the problem of whether the card will be purchased for the reserve price or for something higher. If it is auctioned at the reserve price, the worst that could happen is if the card was auctioned at Vintage, and it would cost the buyer thirty points. This number is called a *column maximum* (it is the maximum of all the numbers in the first column). If the card is purchased for a high price, the column maximum is a hundred points. From the standpoint of potential buyers, who do not know whether the card will go for a reserve price or a high price, the "best of the worsts," the smaller of the two column maximums, is thirty points. This number is called the *minmax*, which is an abbreviation for the minimum of the column maximums. The minmax of thirty points occurs when the card goes for the reserve price.

Originally, game theory was devised under the assumption that two intelligent rivals each could choose a particular strategy. The contestants were simply described as Red and Blue, and the strategies as Red 1 and Red 2, Blue 1 and Blue 2. Each side could view the diagram and decide which strategy to choose, much as the two contestants in rock–paper–scissors are free to make their choices. Dolores is in such a position; she is free to sell her card at either auction house.

In this case, Dolores faces is not a flesh-and-blood rival but a situation. She has no way of knowing whether the card will sell for a high price or a reserve one. However, if she looks at the game as if the world were choosing the best way to thwart her desires, this is the way she would analyze it. The world knows that Dolores' maxmin calls for her to auction the card at Vintage, and she knows that its minmax occurs if the card goes for a reserve price. Even if she tells them she will auction the card at Vintage, she cannot be prevented from making at least $30,000. Even if the world lets us know that the card is going for the reserve price, she cannot make more than $30,000. This number, $30,000, is the *value* of the game.

You may be a little worried that the real world does not conspire in such a fashion as to do its best to ensure that the card will sell at the reserve price. That's certainly true. Nonetheless, there's no way she can tell—the card may not interest any buyer. Markets for collectibles are frequently misjudged. Dolores has to choose her strategy based on the assumption that she has a rational opponent (that was one of the original assumptions of game theory)—and she will accrue added profit if her opponent deviates from the best strategy. No one ever does well in a game by depending on one's opponent to make a mistake. You can hope your opponent makes a mistake, but it's better to plan on the assumption that he or she won't.

Suppose that each side plays its best strategy: Dolores auctioning the card at Vintage and the buyers purchasing it for the reserve price. If either side deviates while the other side plays its best strategy, the side that deviates loses. If Dolores sells at Classic for the reserve price, she loses by doing so. If the card commands a high price at Vintage, Dolores gains (and the world of buyers loses) as a result. This is characteristic of games in which the maxmin is equal to the minmax; the side that deviates from its best strategy loses.

When the maxmin and the minmax have the same value, the game is said to have a *saddle point* (the rider of a horse sits at a place that is simultaneously a maximum and a minimum—the lowest point from the back of the horse to the front, and the highest point from the left side of the horse to the right). In this case, each player elects to follow the single strategy (called a *pure strategy*) that guarantees that the maxmin and minmax are both achieved, and the value of the game is always the maxmin (or the minmax, depending on which of the two opponents you are).

Mixed Strategies

Now let's look at the situation that Freddy encountered. The diagram is repeated below for convenience, but the analysis will be a little clearer if we just look at the classic situation in which there are two intelligent opponents, Red and Blue, each of whom has a choice of two strategies. Remember that high scores are good for Red, low scores are good for Blue.

		Blue	
		1	2
Red	1	10	0
	2	2	7

The first row minimum is 0, the second is 2, so the maxmin is 2. The first column maximum is 10, the second is 7, so the minmax is 7. Let's suppose that Red elects Strategy 2 in order to choose his maxmin strategy and Blue also elects Strategy 2, choosing his minmax strategy. The payoff is 7 points—but Blue can improve his score by switching from Strategy 2 to Strategy 1 if Red sticks with Strategy 2. In fact, no matter which of the four possible choices are made (Red 1 vs. Blue 2, etc.), one side *always* benefits from deviating if the other side continues to play the same strategy. This is the type of thing that's seen in repeated plays of rock–paper–scissors; one side always benefits from switching its strategy if the other side continues to play the same strategy.

This type of situation was analyzed in the story when Freddy was trying to decide whether or not to call Lisa. If there is no saddle point, each side's best long-term procedure is to adopt a strategy whose expectation is the same against either of its opponent's options. Freddy played the role of Red in the story, so let's suppose that Red elects to play Strategy 1 randomly with probability p, and Strategy 2 randomly with probability $1 - p$. We can compute Red's expectation against either of Blue's strategies.

Against Blue 1, Red wins 10 points with probability p and 2 points with probability of $1 - p$, for an expectation of $10p + 2(1 - p)$ $= 8p + 2$.

Against Blue 2, Red wins 0 points with probability p and 7 points with probability of $1 - p$, for an expectation of $0p + 7(1 - p)$ $= 7 - 7p$.

If Red equates these two expectations, nothing Blue can do will prevent Red in the long run from achieving the equated expectation. Solving $8p + 2 = 7 - 7p$ gives $p = 1/3$. Against Blue 1, Red therefore has an expectation of $8 \times 1/3 + 2 = 4^2/3$ points, and against Blue 2, Red has an expectation of $7 - 7 \times 1/3 = 4^2/3$ points, just as in the story.

Analyzing 2 × 2 Games

Step 1: Find the minmax and the maxmin. If these two are equal, the game has a saddle point, and each side should play a pure strategy. The row player should play the strategy with the higher minimum, and the column player should play the strategy with the lower maximum. The value of the game is the maxmin (or the minmax, as they are the same).

Step 2: If the game does not have a saddle point, assume that the row player plays one row with probability p and the other with probability $1 - p$. Compute the expectation of this strategy against each of the column player's two strategies. Equate these two expectations to determine the value of p. The value of the game can be determined by computing the expectation against either strategy using the value of p just determined.

The column player should go through a similar analysis.

APPENDIX 13

ELECTIONS IN "DIVISION OF LABOR"

It has long been a dream of social scientists to come up with an ideal system for translating individual preferences to the preferences of the society. Arrow's Theorem, which was proved by Kenneth Arrow while he was still a grad student, shows that this cannot be accomplished.

VOTING METHODS AND ARROW'S THEOREM

/ (Voting methods continued from p. 120)
You saw three different voting schemes in the Bankers Club Election. Assume that every voter submits a ballot that has a preference listing among all the candidates. For example, if a voter submits a ballot that has Ackroyd as the first choice, Williams as the second, and Morris as the third, then if Ackroyd is not elected, the voter prefers Williams to Morris. Ties are not permitted.

For convenience, the result of the Bankers Club election is reprinted below.

First Choice	Second Choice	Third Choice	Number of Ballots
Ackroyd	Morris	Williams	24
Williams	Morris	Ackroyd	18
Morris	Williams	Ackroyd	12

Election Scheme 1: *Plurality*. The candidate who receives the most first-place vote wins.

In the story, this was Forrest Ackroyd's preferred voting method. This has the advantage that it is generally easy to compute and rarely results in ties (especially if there are a large number of voters). The disadvantage of this scheme is that it is possible for the winner to be loved by a minority of the electorate and vigorously detested by everyone else. It appears that this is how the electorate at the Bankers Club felt about Ackroyd.

Election Scheme 2: *Runoff*. Eliminate all candidates except those who have the two highest first-place totals and then have a secondary election between them.

This was how Helen Williams felt the election should be decided. One of the difficulties with this method is that it lends itself to *insincere voting*. An example of insincere voting could be seen in a four-person race decided by the runoff method, in which there is a clear front-runner, a close contest for the second slot, and a splinter candidate. The splinter candidate could wield a lot of clout by telling his or her supporters to throw their votes to one of the two candidates who are in contention for the second position on the post-runoff ballot. This frequently happens in the real world. *Sincere voting* occurs when each individual lists his or her preferences, letting the chips fall where they may.

Election Scheme 3: *Borda count*. An arithmetical weighting scheme is devised for first place, second place, third place, . . . , with more points given for higher placings (first place gets more points than second, etc.). The winner has the highest total (or highest average per voter).

In its favor, Borda counts reflect how each voter feels about each candidate. One disadvantage of Borda counts is that different Borda counting schemes can result in different winners!

Example 1: Suppose that there are three candidates in an election, whom we shall call A, B, and C, and the balloting produces the following results:

First Choice	Second Choice	Third Choice	Number of Ballots
A	B	C	11
B	C	A	8
C	B	A	17

Compute the results of the election by the Borda count method (a) if the weighting scheme is 3–2–1, and (b) if the weighting scheme is 5–3–2.

Solution: This is basically simple arithmetic. With a 3–2–1 weighting scheme, the point totals are

$$A = 11 \times 3 + 8 \times 1 + 17 \times 1 = 58$$
$$B = 11 \times 2 + 8 \times 3 + 17 \times 2 = 80$$
$$C = 11 \times 1 + 8 \times 2 + 17 \times 3 = 78$$

So B is the winner. However, with a 5–3–2 weighting scheme, the point totals are

$$A = 11 \times 5 + 8 \times 2 + 17 \times 2 = 105$$
$$B = 11 \times 3 + 8 \times 5 + 17 \times 3 = 124$$
$$C = 11 \times 2 + 8 \times 3 + 17 \times 5 = 131$$

In this case, C is the winner. ∎

The Paradox of Transitivity

It has been known for at least a century that difficulties can arise in certain situations. One obvious property that individual preferences display is that, if the individual prefers alternative A to alternative B, and that same individual also prefers alternative B to alternative C, then that individual must prefer alternative A to alternative C. This is called *transitivity*. Numbers display a similar type of transitivity: if $a > b$ and $b > c$, then it follows that $a > c$.

However, transitivity of individual preferences does *not* produce transitivity of the preferences of a majority of society! Look at the following example.

Example 2: Suppose that A, B, and C are candidates in an election and that the voters cast their ballots as follows:

First Choice	Second Choice	Third Choice	Number of Ballots
A	B	C	11
B	C	A	10
C	A	B	9

Notice that A is preferred to B by a majority of voters (20 to 10), and similarly B is preferred to C by a majority of voters (21 to 9). If an individual exhibited these preferences (A over B and B over C), we would be justified in concluding that he or she preferred A to C. However, when a majority prefers A to B and B to C, we cannot reach the same conclusion, for in this example, a majority of voters prefers C to A (by 19 to 11).

This example simply illustrates that there is something wrong with our intuitive ideas about how the preferences of the majority can be deduced from the preferences of individuals.

Arrow's Impossibility Theorem

Example 2, which shows that majority preference is not transitive, is a little unsettling. Nonetheless, it did not dissuade generations of social scientists from seeking a system that would translate the preferences of individuals into preferences for the society.

Ideally, we would like to take a list of individual preferences and from this arrive at a list of the preferences of society. Example 2 shows that we cannot expect transitivity to hold for society's preferences, even though it will certainly hold for the preferences of the individual. We would like to construct a "social preference method" derived from a list of individual preferences that enables society to choose between any two alternatives. Here is a list of some properties, each of which is desirable.

Transitivity: We would like this "social preference method" to be transitive: If society prefers alternative A to alternative B, and it also prefers alternative B to alternative C, then it should also prefer alternative A to alternative C.

Nondictatoriality: We would like a society that is nondictato-rial. In other words, there should be no individual whose preferences are automatically adopted by society. In a dic-tatorship, the dictator's preferences are automatically ad-opted by society; that's what makes a dictator.

Preservation of unanimous preferences: If every member of the society prefers alternative A to alternative B, then the society should prefer alternative A to alternative B.

Independence of irrelevant alternatives: Suppose that the bal-lot contains at least three alternatives, A, B, and C, and that society prefers alternative A to alternative B. Now suppose that alternative C is eliminated from the ballot. Society should still prefer alternative A to alternative B.

Like transitivity, this is generally obvious for individuals. Let's suppose that you are going out for dinner and that steak, fish, chicken, and hamburger are on the menu. You select steak. The waiter comes back and tells you that they ran out of fish. Your reaction would undoubtedly be, "Who cares? Bring me my steak!" The absence of fish is an irrelevant alternative; it would only be relevant if you had actually decided to have fish for dinner.

Each of these properties is not only desirable but also seems ostensibly reasonable. However, Arrow's Impossibility Theorem shows that one cannot construct a social preference method with all of the above properties!

Every time an election involves more than two candidates, there is the possibility that the choice of method may play a critical role in deciding the election, and that the "will of the people" may be inadvertently or unknowingly subverted. Arrow's Theorem shows that there can be no perfect method.

Here's some food for thought: Suppose that the United States had only four states, three of which had twenty-six electoral votes and one of which had twenty-two electoral votes, a total of a hun-dred electoral votes. Too bad for you if you live in the state with twenty-two electoral votes because this state has absolutely no in-fluence on the outcome of the election! If a candidate wins any two of the twenty-six electoral vote states, that's fifty-two electoral votes—all they need.

So is the Electoral College a good idea? As a well-known television network says, we report, you decide—and you may have to, because there are movements afoot to get rid of the Electoral College, which has led to some strange results in presidential elections.

As Arrow's Theorem shows, there's no perfect way to run an election, but social scientists are continually on the lookout for ways that have as few flaws as possible.

APPENDIX 14

ALGORITHMS, EFFICIENCY, AND COMPLEXITY IN "THE QUARTERBACK CONTROVERSY"

THE VALUE OF PLANNING

Going places and doing things inefficiently wastes precious resources. Some of these resources, such as time and money, can be expressed in terms of numbers—and anything that can be expressed in numbers is fair game for analysis.

You're almost certainly familiar with some of the basic principles of operations research, the branch of mathematics concerned with efficient planning, from your everyday life—often, they're just common sense. For instance, if you need to mail a package at the post office, pick up a book at the library, and leave the car to have the oil changed, and all of these locations are within walking distance of one another, you will find out when the library and post office are open so you can mail the package and pick up the book while the oil is being changed. Also, if you have several different locations to visit, you'll try to visit them in an order that minimizes the time you spend getting from one place to another.

In each day, you have only a few things to do and a few places to go. It is therefore fairly easy to plan to do these things with some efficiency because there are only a limited number of ways to do them. However, when there are many things to do and many places to go, the number of possible ways to do them can be astronomically large. For example, in any major construction project, there are a huge number of tasks to be done. A lot of time and money can be wasted if the electricians are sitting around waiting for the wiring conduits to be installed.

In this chapter, we'll look at some of the mathematics of going places (routes) and doing it as efficiently as possible.

GOING PLACES (ROUTES)

All cities, whether large or small, face two common problems involving routes: garbage collection, and the repair of broken traffic lights. A garbage truck starts out from a central location, picks up the garbage at all the buildings on a number of streets, and returns to its starting point. The route is most efficiently designed if the truck does not have to retrace any of the streets.

A traffic light repair crew faces a different problem. Broken traffic lights undoubtedly occur at different areas of the city, and the route for repairing them is most efficiently designed if the total distance the repair crew has to travel is kept as small as possible.

Each of the routing problems can be presented in the simplified form of a *graph*, as indicated below (yes, the word "graph" has a different meaning when discussing functions, but mathematicians

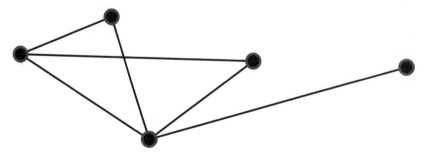

Figure 14.2. A Graph with Five Vertices and Six Edges

are not the only ones to assign multiple meanings to the same word—look up "spring" in the dictionary!). A graph consists of points, called *vertices*, connected by lines, called *edges*.

The garbage collection problem is to design routes that retrace as few edges as possible. The traffic light repair problem is to minimize the total distance traveled in visiting all the vertices.

When the Goal Is to Visit All the Vertices

As you saw in "The Quarterback Controversy," the problem of visiting all the vertices in a graph while minimizing the total distance traveled is known as the Traveling Salesman Problem, abbreviated TSP.

If you assume that a salesman (or woman) has to visit n different cities before returning home, he (or she) has a choice of n different cities to visit first. From there, he (or she) could go to any one of the $(n-1)$ unvisited cities, and from that city to any one of the $(n-2)$ unvisited cities, etc. By the Chinese Restaurant Principle, there are a total of $n \times (n-1) \times (n-2) \times \ldots \times 1 = n!$ different routes that he (or she) could take.

You may recall from the chapter on counting that, even for fairly small values of n, such as 25, $n!$ is an astronomically large number. Even the fastest supercomputer would take billions of years to examine the total distances of each of 25! different routes. As a result, mathematicians have examined two different questions.

> *Question 1*: Is there an algorithm that enables one to examine only a select handful (this can be defined in mathematical terms, but don't worry about it) of routes and still come up with the shortest route?
>
> *Question 2*: Is there an algorithm that enables one to examine only a select handful of routes and come close to the shortest route?

(Traveling Salesman Problem continued from p. 125)
As Pete points out in the story, as far as question 1 is concerned, no one even knows whether such an algorithm exists, although the betting in the mathematical community is that it doesn't. The

Traveling Salesman Problem is an example of what mathematicians call an *NP-complete* problem. There are many extremely important such problems, and they generally involve a factorial number of possibilities.

Consider, for instance, a scheduling problem that might take place in a typical factory, such as the problem of assembling a car, a TV, or a circuit board. Many different subtasks have to be performed, and although some must clearly follow others, it is often possible to perform many of the subtasks in any order. Here it is necessary to minimize the total time, or possibly the total cost, of performing the entire job, but it is just the TSP in another guise.

Any problem that involves factorials is troublesome because a problem with "factorially many" computations to make requires far too many computations even for fairly small numbers. A traveling salesman visiting 25 cities is way beyond the power of even the fastest supercomputer to handle, and variations of the TSP often have the equivalent of thousands of cities.

In 1971, Stephen Cook, a mathematician at the University of Toronto, showed that, if one NP-complete problem could be solved, they could all be solved. A result such as this does not tell *how* to solve a problem, but it yields a certain amount of insight. Additionally, it shows that (1) if someone can solve just one of these problems, they can all be solved, and (2) if someone can show that just one of these problems *cannot* be solved, there is no need to "waste time" trying to solve any of the others. To date, no one has solved any NP-complete problem, but no one has shown that they cannot be solved, either.

More progress has been made in answering question 2. It is fairly easy to describe an algorithm, such as the "nearest neighbor" algorithm that appears in the story, and to execute it. What is substantially more difficult is to figure out how good that algorithm is. Obviously, any TSP has a "best" answer—the number of miles of the shortest route. If one could find an algorithm that guaranteed an answer within, say, 10% of the best answer, this would obviously represent substantial progress. Several algorithms have been devised that give excellent results with problems that occur in the real world, but no algorithm is yet known that guarantees "coming close" in all cases.

Because of the immense practical value of the TSP and its re-lated NP-complete problems, this is one of the most intensively investigated of all mathematical problems.

Calculating Task Complexity

A task that is doable in N^2 steps, or N^8 steps, or N^p steps for any fixed integer p is said to be a *polynomially complex task*. Polyno-mially complex tasks have the following property: The price tag for increasing the number N by "just 1 more" gets smaller and smaller as N gets larger and larger.

Consider a task that is doable in N^3 steps. If $N = 10$, then the task requires 1,000 steps. If $N = 11$, the task requires 1,331 steps, an increase of about 13.3%. However, if $N = 100$, the task requires 1,000,000 steps, but if $N = 101$, the task requires 1,030,301 steps, an increase of only about 3%. This "price tag" gets smaller as N gets larger.

The next stage of task magnitude is the *exponentially com-plex task*, which might require 2^N steps to complete. For such a task, the price tag of "just one more" is always the same—the task time doubles. Obviously, any algorithm that can reduce an exponentially complex task to a polynomially complex task rep-resents a tremendous potential savings in time, especially when N is large.

The ultimate horror show in task complexity is the *factorially complex task*, such as the TSP. As you have seen, a traveling sales-man who must visit N cities has a choice of $N!$ possible routes. The cost of "just one more" for a factorially complex task gets worse and worse as N gets larger and larger. In fact, since $(N + 1)!/N! = N + 1$, you can see that the cost of going from N to $N + 1$ increases by an ever-increasing factor of $N + 1$.

The "nearest neighbor" algorithm discussed in "The Quarter-back Controversy" reduces a factorially complex problem to a polynomially complex one. Applying the "nearest neighbor" algo-rithm to an N-city TSP requires one simply to look at N numbers for the first intercity trip, then $N - 1$ numbers for the second inter-city trip, $N - 2$ numbers for the third intercity trip, and so on. This method would give a total of $N + (N - 1) + \ldots + 1$ computational steps, and we know (from the chapter on patterns) that this total is $N \times (N + 1)/2$, which is less than N^2.

The "nearest neighbor" algorithm is often called a *greedy* algorithm because it decides at each stage what is best according to a certain rule and then hopes that this step-by-step plan gives the best overall result. This is somewhat akin to an individual who eats the first item of food he (or she) sees whenever he (or she) is hungry and hopes that by so doing his (or her) nutritional needs will be best satisfied. Good luck with that method.

AN INTRODUCTION
TO SPORTS BETTING

Betting on sports almost certainly started thousands of years ago, when one horse owner said to another, "My horse is faster than yours," and the second responded, "Oh, yeah? Wanna bet?" Since then, betting on sports has become an avocation of millions of people and is a billion-dollar industry that shares many of the ideas and techniques of two trillion-dollar industries: financial markets and insurance. Different societies have different views on sports betting; it is a respectable national institution in England, but less so in the United States, where the laws governing sports betting vary from state to state. Offshore betting websites proliferate; some abide by federal restrictions, and others don't. The rule "Buyer beware!" is good advice to anyone interested in patronizing such a website.

PARI-MUTUEL BETTING

Legalized horse racing is common throughout the world, and the usual method of wagering is pari-mutuel betting. Bettors place bets on which horse will win; all these bets are placed in a pool. The organization supervising the wagering collects a fraction of the pool and pays the remaining money back to the winners.

Here's an easy example. Suppose that there are three horses in a race, and the proprietors supervising the wagering take 20% of the total amount bet (this is a fairly typical percentage in state-run horse racing). The money has been wagered as follows

Horse	Amount Bet to Win on That Horse
Alexander the Great	$500
Mike the Mediocre	$400
Howard the Horrible	$100

The total wagered is $1,000. So 20% of that, or $200, is deducted by the proprietors, leaving $800 to be distributed among the winners. If Alexander the Great wins, each dollar bet on Alexander the Great returns $800/500 = $1.60. In this case, a bettor who bet $1 on Alexander will have paid $1 for a ticket; he or she will then receive $1.60, a profit of $0.60 on his or her bet. If Mike the Mediocre wins, each dollar bet on Mike the Mediocre returns $800/$400 = $2.00; a profit of $1.00. If Howard the Horrible wins, each dollar bet on Howard returns $800/$100 = $8.00; a profit of $7.00.

LINE BETTING

Many sports events are a contest between two teams or individuals. As an example, let's suppose that the Dallas Cowboys are playing the New York Giants at New York. New York is felt by the Las Vegas bookmakers to be a slightly stronger team than Dallas, and the fact that the game is being played at New York gives the Giants a home-field advantage, which is normally felt to be worth three points. That, combined with the fact that the Giants are felt to be a slightly stronger team, leads the Las Vegas bookmakers to estimate that, on average, the Giants should win by five points. The Las Vegas bookmakers therefore set the line at New York minus five. A bettor who bets on the Giants is said to give five points, and the Giants are known as the *favorite*. A bettor who bets on the Cowboys is said to get five points, and the Cowboys are known as the *underdog*, commonly referred to as the *dog*.

When the game ends, if New York wins by MORE than five points, those people who have bet on New York win the bet, those who have bet on Dallas lose the bet. In this case, the favorite is said to have *covered*. If New York either loses or wins by LESS than five points, those people who have bet on Dallas win the bet, and those who have bet on New York lose the bet. If New York wins by EXACTLY five points, the game is said to be a *push*, and no money changes hands.

If you bet with a bookmaker, you generally must pay odds of 11 to 10 on a losing bet. For instance, if the line is New York minus five (or Dallas plus five, which is the same thing), and you bet $100, if you win you are paid $100. However, if you lose, you must pay $110. The extra $10 paid on losing bets is known as *the vig*, which is short for *vigorish*. The term originated in the twentieth century from the Russian *vyigrish* (brought over from Russia and incorporated into Yiddish), meaning "gains" or "winnings."

NOTES

CHAPTER 8. NOTHING TO CROW ABOUT

1. *Handicapping* is the process of predicting the success of the various entrants in a contest. Political pundits often handicap various elections.

2. A *longshot* in an athletic contest is a contestant with little chance to win. In everyday English, it refers to an improbable event.

3. In poker, a *hole card* is a card dealt to a player that only that player can see. Cards that can be seen by all players are called *up cards*; they are dealt face up. The expression *ace in the hole* means a fact known only to its possessor that can be extremely advantageous. During World War II, the fact that the British had developed radar in secret was an ace in the hole.

4. In horse racing, the *daily double* is a bet on which horses will win the first two races. A daily double bet selects one horse in the first race and one horse in the second race, and the bet wins only if both horses win.

5. In horse racing, to *wheel* a particular selection means to place bets on that selection and all the other horses in the race. For example, if a bettor is sure that Alexander the Great will win the first race, he or she might buy a number of daily double tickets by wheeling Alexander the Great with every horse in the second race. If Alexander the Great actually wins and a longshot wins the second race, this method can be extremely profitable.

CHAPTER 9. THE WINNING STREAK

1. One of the common problems confronting both a sports bettor and an investor is how to allocate funds to investments with differing degrees of risk. A detailed analysis of this problem, known as portfolio theory, won the economist Harry Markowitz the 1990 Nobel Prize in economics. Bettors generally solve this problem by deciding on a basic unit for each bet—such as $100. If their analysis shows that a particular bet is unusually attractive, they might make a larger bet. If the basic unit is $100, a $200 bet is known as a *two-unit bet*, a $300 bet as a *three-unit bet*.

CHAPTER 10. ONE LONG SEASON

1. When a betting line includes a half point, such as three and a half, a push is no longer possible, since the final result of a game is always a whole number. If the line is three and a half, the favorite covers if it wins by four or more.

2. An *if bet* is a second bet that is contingent on winning a first bet. Pete has bet $200 on the Knicks and has told his bookie that if the Knicks cover and the line on the Lakers is six points or better, the $200 that he won on the Knicks should then be bet on the Lakers.

CHAPTER 11. THE GREAT BASKETBALL FIX

1. Many bookmakers offer a separate line for the total number of points scored in a game. This is known as the *totals line*, and the bettor selects either *over*, betting that the total number of points scored by both teams will be more than the totals line, or *under*, betting that the total number of points scored by both teams will be less than the totals line. Winning bets are paid off at even money, losing ones pay off at 11–10, just like the points line.

2. The process of hedging is common in investment; it is an attempt to obtain a profit no matter what the outcome of the event. If a bookmaker offers a line such as minus five, the ideal situation for the bookmaker would be to have the same amount bet on the favorite and the dog. That way, unless the game is a push, he would win 110% of the money on the bets he or she won than he or she lost on the losing bets—guaranteeing a profit. However, sometimes a line attracts a preponderance of betting on one side. A bookmaker may offer a line and find out that four times as much is bet on the favorite as on the dog. In that case, the bookmaker can increase the line to attract more money on the dog.

Here's a simple example.

		$ Bet on Favorite	$ Bet on Dog
Original line	Favorite minus 5½	4,000	1,000
Modified line	Favorite minus 6½	1,000	4,000

All is well for the bookmaker as long as the favorite does not win by precisely six points. For instance, if the favorite wins by seven, the bookmaker loses $4,000 to those who bet on the favorite at the original line and wins $1,100 from those who bet on the dog at the original line. He also loses $1,000 to those who bet on the favorite at the modified line but wins $4,400 from those who bet on the dog at the modified line. This results in a net profit of $500 to the bookmaker.

However, catastrophe strikes the bookmaker if the game ends at six points, in the middle of those two lines. He wins $1,100 from those who bet on the dog at the original line and also $1,100 from those who bet on the favorite at the modified line. However, he loses $4,000 to those who bet on the favorite at the original line and another $4,000 to those who bet on the dog at the modified line, for a net loss of $5,800. In effect, the bookmaker is offering odds of $5,800 to $500 that the game will not end with the favorite winning by six. In this case, the bookmaker is said to have been *middled*. In the above example, if the modified line had been Favorite minus 8½ (which might have happened had 6½ not attracted enough betting on the dog), the bookmaker would be middled if the game ended with the favorite winning by six, seven, or eight. The larger the gap between the prices, the greater the chance that the bookmaker will be middled.

This is also a problem in the commodities futures markets, which operate in a fashion that is greatly similar to sports betting. In this case, the line is known as the futures price of the commodity, and the result of the "game" is the price of the commodity at a particular future point in time, known as the settlement price. The major difference is that the value of a winning or losing "bet" in the commodities futures market is not a fixed amount (such as a $100 bet) but the difference between the price at which the futures were bought and the price at which the commodity settles, much as the value of a stock purchase depends upon the difference between the price at which the stock is bought and the price at which it is sold.

INDEX